T0213987

Key 5G Physical Layer Technologies

Douglas H. Morais

Key 5G Physical Layer Technologies

Enabling Mobile and Fixed Wireless Access

 Springer

Douglas H. Morais
Wireless Communications Consultant
San Mateo, CA, USA

Contents from this book were previously published by Pearson Education, Inc.

ISBN 978-3-030-51443-3   ISBN 978-3-030-51441-9   (eBook)
https://doi.org/10.1007/978-3-030-51441-9

This Springer imprint is published by the registered company Springer Nature Switzerland AG
The registered company address is: Gewerbestrasse 11, 6330 Cham, Switzerland

To my wife Christiane,
and our grandchildren
Ruairi, Aideen, and Felix
and
In loving memory of my "big brother" Peter

Preface

The Third Generation Partnership Project's (3GPP's) Fifth Generation (5G) mobile technology represents a significant advance in capability and applicability over its 4G predecessor, Long Term Evolution (LTE). That said, it continues on the evolutionary path, in that a number (but not all) of the key technologies that it utilizes are inherited from LTE. The complete 5G specified system encompasses a wide architecture from the core network to the end user. This text addresses primarily the 5G *Physical Layer* and more specifically those key technologies that facilitate the functioning of the physical layer. The physical layer, as the name implies, embodies those functions related to the physical realization of transmission between a base station and end user equipment. Among the physical layer technologies covered here are digital modulation, channel coding, multi-carrier based multiple-access techniques, multiple antenna techniques including beamforming, and a number of others. Following the presentation of these technologies, a high-level description of 3GPP's new 5G radio access system known as New Radio (NR) is given and it is demonstrated how the key technologies presented earlier facilitate the attainment of many of the goals of the NR standards, including very low latency and the transmission of very high-speed data. Because of this very high data rate capability, NR synergistically supports competitive, very high data rate, fixed wireless access (FWA). As there is little to date in textbook literature on 5G FWA, this text fills the vacuum by devoting the last chapter to a high-level overview and the technical physical layer aspects of this important and expanding wireless application.

Several texts already exist that cover, individually, some of the key technologies addressed in this book, in addition to a multitude of others. No single one, however, as best the author can ascertain, addresses all the physical layer technologies covered here, all of which the author views to be key. Further, most, in general, cover technologies at a level that presupposes that the reader is already familiar with those technologies and seeks a deeper understanding. This book, on the other hand, presupposes only a general technical background in telecommunications and possibly no knowledge of some or all of the specific technologies covered.

The material presented is directed to industry professionals as well as academics. On the industry side, it should prove valuable to engineering managers, system engineers, technicians, and anyone who would benefit from a rounded understanding of the key physical layer technologies of 3GPP's 5G NR in order to more effectively execute their job. On the academic side, it should provide the upper undergraduate or graduate university student with a useful introduction to the physical layer aspects of 5G NR. The material presented is intentionally not overly rigorous so as to be friendly to a wide audience and in keeping with a desire to convey a somewhat high-level view of the presented technologies. Mathematics, though clearly necessary for any meaningful study of the subject at hand, has been minimized. Further, a chapter on relevant mathematical concepts applied throughout the text has been included for the benefit of those to whom it may be unfamiliar or those who would benefit from a refresher course. In addition, mathematical formulae applied in certain derivations in the text is provided in the Appendix. For those desiring to explore some or all of the material presented in greater detail, several references are provided. Further, a number of texts currently exist, and more will no doubt continue to be released, that collectively treat the technologies presented here in greater detail. A goal of this text is to allow the reader, if he/she so desires, to tackle these more advanced books with confidence.

San Mateo, CA, USA   Douglas H. Morais

Abbreviations and Acronyms

3GPP	Third Generation Partnership Project
4G	Fourth Generation
5G	Fifth Generation
A/D	Analog to Digital
ABPSK	Associated Binary Phase Shift Keying
ACK	Acknowledgement
ADC	Analog to Digital Converter
AMC	Adaptive Modulation and Coding
ARQ	Automatic Repeat Request
BCCH	Broadcast Control Channel
BCH	Broadcast Channel
BER	Bit Error Rate
BLER	Block Level Error Rate
BPSK	Binary Phase Shift Keying
BS	Base Station
CA	Carrier Aggregation
CCCH	Common Control Channel
CDD	Cyclic Delay Diversity
CDF	Cumulative Distribution Function
CFO	Carrier frequency Offset
CN	Check Node
CoMP	Coordinated Multipoint Transmission/Reception
CP	Cyclic Prefix
CPE	Common Phase Error
CPE	Customer Premise Equipment
CRC	Cyclic Redundancy Check
CSI	Channel State information
CSI-RS	Channel-State Information Reference Signal
D/A	Digital to Analog
DAC	Digital to Analog Converter
DCCH	Downlink Control Channel

DCI   Downlink Control Information
DFDMA   Distributed Frequency Division Multiple Access
DFT   Discrete Fourier Transform
DL   Downlink
DL-SCH   Downlink Shared Channel
DMIMO   Distributed MIMO
DM-RS   Demodulation Reference Signal
DRB   Data Radio Bearer
DSBSC   Double Sideband Suppressed Carrier
DTCH   Dedicated Traffic Channel
EIRP   Equivalent Isotropically Radiated Power
eMBB   enhanced Mobile Broadband
EPC   Evolved Packet Core
ETSI   European Telecommunications Standards Institute
FBMC   Filter Bank Multi-Carrier
FDD   Frequency Division Duplexing
FDM   Frequency Division Multiplexing
FD-MIMO   Full Dimension MIMO
FEC   Forward Error Correction
FFR   Fractional Frequency Reuse
FFT   Fast Fourier Transform
FSDD   Frequency Switched Division Multiplexing
FWA   Fixed Wireless Access
GFDM   Generalized Frequency Division Multiplexing
GSM   Global System for Mobile Communications
HARQ   Hybrid Automatic Repeat Request
H-FDD   Half Duplex-Frequency Division Duplexing
ICI   Inter-Carrier Interference
IDFT   Inverse Discrete Fourier Transform
IFDMA   Interleaved Frequency Division Multiple Access
IFFT   Inverse Fast Fourier Transform
IoT   Internet of Things
IP   Internet Protocol
ISD   Inter-cell Site Distance
ISI   Inter Symbol Interference
ITU   International Telecommunications Union
LDPC   Low Density Parity Check
LFDMA   Localized Frequency Division Multiple Access
LLR   Log Likelihood Ratio
LNA   Low Noise Amplifier
LOS   Line-of-Sight
LTE   Long Term Evolution
MAC   Medium Access Control
MBB   Mobile Broadband
MIB   Master Information Block

MIMO	Multiple-Input Multiple-Output
MISO	Multiple-Input single-Output
mMIMO	massive MIMO
mMTC	massive Machine Type Communications
MRC	Maximum Ratio Combiner
MU	Mobile Unit
MU-MOMO	Multi-User MIMO
NACK	Negative Acknowledgement
NLOS	Non Line-of-Sight
NOMA	Non Orthogonal Multiple Access
NR	New Radio
OFDM	Orthogonal Frequency Division Multiplexing
OFDMA	Orthogonal Frequency Division Multiplex Access
OOB	Out of Band
OQAM	Offset Quadrature Amplitude Modulation
P/S	Parallel to Serial
PAM	Pulse Amplitude Modulation
PAPR	Peak-to-Average Power Ratio
PBCH	Physical Broadband Channel
PCCH	Paging Control Channel
PCH	Paging Channel
PCM	Parity Check Matrix
PDCCH	Physical Downlink Control Channel
PDCP	Packet Data Convergence Protocol
PDF	Probability Distribution function
PDSCH	Physical Downlink Shared Channel
PDU	Protocol Data Unit
PMP	Point-to-Multipoint
PN	Phase Noise
PRACH	Physical Random-Access Channel
PRB	Physical Resource Block
PSD	Probability Spectral Density
PSS	Primary Synchronization Signal
PT-RS	Phase-Tracking reference Signal
PUCCH	Physical Uplink Control Channel
PUSCH	Physical Uplink Shared Channel
QAM	Quadrature Amplitude Modulation
QoS	Quality of Service
QPSK	Quadrature Phase Shift Keying
RACH	Random Access Channel
RAN	Radio Access Network
RB	Resource Block
RF	Radio Frequency
RLC	Radio Link Control
RMS	Root Mean Squared

RMSI	Remaining Minimum System Information
RNTI	Radio Network Temporary Identifier
ROHC	Robust Header Compression
RRC	Radio Resource Control
RRC	Root Raised Cosine
S/P	Serial to Parallel
SC	Successive Cancellation
SCL	Successive Cancellation List
SCS	Sub-carrier Spacing
SD	Space Diversity
SDAP	Service Data Adaptation Protocol
SDU	Service Data Unit
SFBC	Space-Frequency Block Code
SI	System Information
SIB	System Information Block
SIC	Successive Interference Cancellation
SIMO	Single-Input Multiple-Output
SINR	Signal to Interference and Noise Ratio
SISO	Single-Input Single-Output
SM	Spatial Multiplexing
SME	Small and Medium Sized Enterprise
SM-MIMO	Spatial Multiplexing MIMO
SNR	Signal to Noise Ratio
SRB	Signal Radio Bearer
SRS	Sounding Reference Signal
SSS	Secondary Synchronization Signal
STBC	Space-Time Block Code
SU-MIMO	Single-User MIMO
TCP	Transmission Control Protocol
TDD	Time Division Duplexing
TRS	Tracking Reference Signal
TTI	Transmission Time Interval
UCI	Uplink Control Information
UDP	User Datagram Protocol
UE	User Equipment
UFMC	Universal Filtered Multi-Carrier
UL	Uplink
UL-SCH	Uplink Shared Channel
UMTS	Universal Mobile Telecommunications System
URLLC	Ultra-Reliable and Low Latency Communications
VN	Variable Node
VoIP	Voice Over Internet Protocol
VRB	Variable Resource Block

Contents

About the Author

Douglas H. Morais has decades of experience in the wireless communications field that encompasses product design, engineering management, executive management, consulting, and short course lecturing. He holds a B.Sc. from the University of Edinburgh, Scotland, an M.Sc. from the University of California, Berkeley, and a Ph.D. from the University of Ottawa, Canada, all in electrical engineering. Additionally, he is a graduate of the AEA/Stanford Executive Institute, Stanford University, California, is a Life Senior member of the IEEE, and a member of the IEEE Communications Society.

Dr. Morais has authored several papers on wireless digital communications; holds three US patents, one on point-to-multipoint wireless communications and two on digital modulation; and has authored the book *Fixed Broadband Wireless Communications* published by Prentice Hall PTR.

Chapter 1
Mobile and Fixed Wireless Cellular Systems Introduction

1.1 Mobile and Fixed Wireless Access Systems

Wireless communications in today's world is diverse and ubiquitous. However, its largest and most visible segment is cellular mobile networks. Originally these networks provided voice communication only over analog networks. However, as outlined below, they have, over time, evolved into all digital networks providing more and more data capacity, to the point that the latest such networks communicate packet switched data only. Now voice, video, and other applications are all converted to data and integrated into a common packet switched data stream. In such mobile networks, the coverage area consists of multiple adjacent cells, with each cell containing a fixed *base station* (BS). Transmission to and from an individual *mobile unit* (MU) is normally between that unit and the base station that provides the best communication. Such communication is commonly referred to as *mobile access*. The actual communication between a BS and a MU takes place over what is defined as the *physical layer*. This layer is responsible for communicating data at an acceptable level of reliability and, when required, at the highest rate possible for the given frequency resources.

Though mobile access networks are designed to communicate with mobile units, nothing prevents them from communicating with fixed units located in homes or small- and medium-sized enterprises (SMEs). In fact, communications of this form are typically of a higher quality than that with mobile units due to the use of higher-gain antennas, possibly higher output power, and the elimination of mobile-induced fading. Such communication is commonly referred to as *fixed wireless access* (FWA).

Though this text outlines, at a high level, the key features of *Fifth-Generation* (5G) mobile systems, it does not strive to provide a comprehensive review of the detailed specifications behind such systems. These details can be easily found elsewhere. Rather, it deals with underlying key technologies that assist in enabling 5G mobile and fixed wireless access not usually covered within the specifications themselves. In particular, it emphasizes those underlying technologies that facilitate

© Springer Nature Switzerland AG 2020
D. H. Morais, *Key 5G Physical Layer Technologies*,
https://doi.org/10.1007/978-3-030-51441-9_1

physical layer transmission, allowing the reader to fully comprehend how very high data rates are achieved and how control of the communication link is established and maintained. In addition, it deals with the application of 5G technology to FWA networks.

1.2   Brief History of Mobile Access

The first commercial cellular telephone systems were introduced in the early 1980s, used analog technology, and are now referred to as *First-Generation* (1G) systems. They were designed primarily for the delivery of voice services.

Second-Generation (2G) systems appeared in the early 1990s. Though, like 1G systems, they were primarily aimed at voice services, they utilized digital modulation, allowing a higher voice capacity and support of low rate data applications. Examples of 2G systems include the European designed *Global System for Mobile Communications* (GSM) system, the American designed IS-136 TDMA and IS-95 CDMA systems, and the Japanese designed PDC system. As an example of data capabilities, the original GSM standard supported circuit switched data at 9.6 kb/s. By the late 1990s, however, with the introduction of *Enhanced Data Rate for GSM Evolution* (EDGE), user rates of between 80 and 120 kb/s were supported.

The process of broadly defining 3G standards for worldwide application was started by the International Telecommunications Union (ITU), which referred to such systems as *International Mobile Telecommunications 2000* (IMT-2000) systems. 3G systems became available in the early 2000s and represented a significant leap over 2G ones. These systems were conceived to deliver a wide range of services, including telephony, higher speed data than available with 2G, video, paging, and messaging.

The European Telecommunications Standard Institute (ETSI) was initially responsible for standardization of an IMT-2000 compliant system to be called *Universal Mobile Telecommunications System* (UMTS) and to be an evolution of GSM. However, in 1998 the *Third-Generation Partnership Project* (3GPP) was created with the mandate to continue this standardization work under the auspices of not only ETSI but also under those of other regional standardization development organizations, thus making the project a more global effort.

At about the same time that 3GPP was being created, a similar organization called 3GPP2 was being created under the auspices of North American, Japanese, and Chinese telecommunication associations. Like 3GPP, its goal was to standardize on an IMT-2000 compliant system.

All the 3G systems standardized by 3GPP and 3GPP2 utilized *Code Division Multiple Access* (CDMA) technology, provided significant increase in voice capacity, and offered much higher data rates over both circuit-switched and packet-switched bearers. Examples of early 3G systems include (a) the first release of UMTS, Release 99, referred to as WCDMA, (b) the first 3GPP2 release called

CDMA 2000, and (c) a later release of 3GPP2 called EV-DO. These systems initially all provided peak *downlink* (DL) and *uplink* (UL) rates in the high kb/s to the very low Mb/s rates. These rates increased over time, but the breakout technology was UMTS whose continued evolution has led to extremely high data rates. Its CDMA-based Release 7, labeled HSPA+, with first service in 2011, provided a maximum DL rate of 84 Mb/s and a maximum UL rate of 23 Mb/s. Under its Release-8 however, it released not only a CDMA-based update but an alternative system based on *orthogonal frequency division multiplexing* (OFDM) technology. This alternative system was labeled *Long-Term Evolution* (LTE). It was designed to permit a DL maximum data rate of 300 Mb/s and an UL maximum data rate of 75 Mb/s and commenced service in 2009.

The ITU commenced work on 4G systems in 2005 and labeled such systems *IMT-Advanced*. Such systems were defined as those capable of providing a maximum DL data rate of 1 Gb/s under low mobility conditions and of 100 Mb/s under high mobility conditions. 3GPP's candidate for IMT-Advanced was *LTE-Advanced*. The IEEE's candidate was 802.16 m called *WirelessMAN-Advanced*. Both these candidates were based on OFDM technology. In October 2010, the ITU announced that LTE-Advanced and WirelessMAN-Advanced had been accorded the official designation of IMT-Advanced, qualifying them as true 4G technologies. However, as the UMTS evolved products had established such an overwhelming market position by that time, WirelessMAN-Advanced failed to take off, leaving LTE-Advanced as the only true 4G technology commercially available, with deployments commencing in the early 2010s.

The ITU in 2015 defined Fifth-Generation (5G) systems as those that meet its IMT-2020 requirements. IMT-2020 envisages the support of many usage scenarios, three of which it identified: enhanced mobile broadband (eMBB), ultra-reliable and low-latency communications (URLLC), and massive machine-type communications (mMTC):

- eMBB is the natural evolution of broadband services provide by 4G networks. It addresses human-centric use cases and applications, enabling the higher data rates and data volumes required to support ever-invcreasing multimedia services.
- URLLC addresses services requiring very low latency and very high reliability. Examples are factory automation, self-driving automobiles, and remote medical surgery.
- mMTC is purely machine centric, addressing services that provide connectivity to a massive number of devices, driven by the growth of the Internet of Things (IoT). Such devices are expected to communicate only sporadically and then only a small amount of data. Thus, support of high data rates here is of less importance.

Among the many IMT-2020 envisaged requirements are (a) the capability of providing a peak DL data rate of 20 Gb/s and a peak UL data rate of 10 Gb/s, (b) user experienced DL data rates of up to 100 Mb/s and UL data rates of up to 50 Mb/s, (c) over the air latency of 1 ms, and (d) operation during mobility of up to 500 km/s.

3GPP commenced standardization works on its 5G wireless access technology in 2016 and labeled it *New Radio* (NR). It decided to address 5G in two phases. The first phase, Release 15 (Rel-15), only addresses eMBB and URLCC. Rel-15 was released in 2019. Rel-16 addresses all IMT-2020 scenarios, meets all key capabilities, and was released in 2020. The first version of Rel-15, ver. 15.1.0, allowed operation in two broad frequency ranges: sub-6 GHz (450 MHz to 6 GHz), referred to as FR1, and millimeter wave (from 24.25 to 52.6 GHz), referred to as FR2. Rel-15, ver. 15.5.0, expanded FR1's range (410 MHz to 7.125 GHz) while leaving the FR2 range unchanged. Networks with performance approaching that of 5G NR Rel-15 specifications began operating in 2019.

1.3   Brief History of Fixed Wireless Access

Fixed wireless access implies a relatively high data rate (tens of Mb/s to 1+ Gb/s) two-way wireless connection between a fixed user location and a transceiver station that is connected to a packet data handling core network. Thus, a single wireless link connecting a fixed user location to the core network via a unidirectional transceiver technically qualifies as an FWA system. In fact, in the 1990s, such links began to be installed. From an economic point of view, however, such systems are rarely cost-effective, and so deployment is limited. If, on the other hand, a base station is used to provide connection to several fixed user locations, the economics become much more attractive as the base station cost is shared by all users. As a result, in a practical sense, FWA today implies access via a point-to-multipoint wireless system.

Early in the 2000s, a market for point-to-multipoint FWA systems seemed attractive. The IEEE embarked on a standardization process for such systems. The result was the 802.16 standard, released in 2002 and covering systems operating between 10 and 66 GHz. This was followed by the 802.16a standard, released in 2003, and covering systems operating between 2 and 11 GHz. Later 802.16a was expanded to address mobile as well as fixed access. All 802.16 standards were commercialized under the name WiMAX. In Europe, meanwhile, two FWA standards were developed by ETSI. One, called HIPERACCESS and similar to 802.16, applied to systems operating above 11 GHz. The other, called HIPERMAN and similar to 802.16a, applied to systems operating below 11GHz. While the abovementioned standards were being developed, entrepreneurial companies began developing and marketing systems based on IEEEs WiFi standard, 802.11. Despite being developed for indoor application, this standard lent itself to outdoor application. Though FWA systems were deployed in the 2000s based on the standards mentioned above or proprietary technology, the market as a whole never really took off. The lack of true economies of scale coupled with relatively low subscriber data rates was the enemy.

This situation began to change in the early 2010s with the advent of 4G technology. Much higher data rates and economies of scale, afforded by piggy backing on 4G mobile systems, led to a reappraisal of the viability of the FWA market (see Fig. 1.1 which depicts a *point-to-multipoint* (PMP) structure where one base station

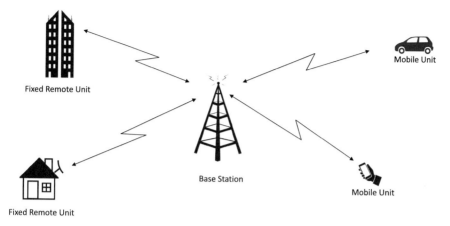

Fig. 1.1 Base station serving both mobile and fixed subscribers

is shown serving both mobile and fixed users). Thus, by the mid-2010s, large companies such as Ericsson and Nokia began offering 4G-based FWA systems overlaid on mobile systems. Data rates competitive with fiber optic cable remained elusive, however, limiting large-scale acceptance of this access approach. The expected game changer is 5G which is expected to provide cable competitive data rates. In fact, one of the earliest applications of 5G technology was FWA, the deployment of such systems having commenced in 2019.

1.4 Base Station: Subscriber Link

Communication links between mobile units, referred to by 3GPP as *user equipment* (UE), and base stations are the most ubiquitous wireless links globally. Every hand-held phone and every vehicle installed voice/data communication device communicates via such a link. Because the transmission path on such a link is or can be constantly changing, designing for reliable communications presents a significant challenge. Path performance can be impacted by operating frequency, terrain variations, atmospheric variations, reflective surfaces, etc. Not surprisingly, therefore, numerous techniques have been devised to optimize reliability. This text reviews many such techniques.

The link between a fixed user location and a base station in a point-to-multipoint FWA system is identical to that between a mobile unit and the base station with the exception that the path does not suffer from fading and distortion as a result of excessive motion. Further, the fixed location equipment can be equipped with a higher-gain antenna relative to the mobile unit leading to better signal strength in both directions and hence higher throughput and reliability. This text reviews the FWA link and the many fixed user location equipment options.

1.5 Cellular Systems Coverage Methods

In this section the main techniques used to effect large, seamless, mobile communications coverage, via what is termed the *cellular concept* [1], are reviewed at a high level to aid in creating a broad view of the 5G operating environment. In the cellular concept, a single very high-power transmitter providing coverage over a very large area is replaced with several lower powered transmitters providing coverage to smaller contiguous areas which collectively provide the same coverage as that provided by the large transmitter. These smaller areas are called cells, hence the nomenclature.

1.5.1 Basic Cellular Coverage

The fundamental transmission structure employed in mobile communications within a cell is to communicate with UEs from a BS in a PMP mode. For planning purposes, cells are normally depicted as hexagons. The hexagonal cell shape is an idealistic model, but it is universally used as it permits manageable analysis. The cellular structure seeks to maximizes capacity for a given spectrum allocation. In the traditional structure, each cell is assigned a portion of the total available spectrum, and its surrounding cells each assigned different but equal portions, resulting in minimization of interference between cells. Normally, in such a structure, omnidirectional antennas are located in the approximate center of the cell. As demand for service increases, individual cell sizes can be decreased, increasing the number of cells covering a given area. Thus, the capacity available to that area is increased, without increase in radio spectrum. Base station antenna radiation is designed to limit coverage to the boundary of the base station's cell. As a result, the same spectrum may be used to cover different cells that are separated from one another by distances large enough to limit interference to tolerable levels.

The process of selecting and allocating spectrum for base stations within a cellular system is called *frequency reuse* [1]. The frequency reuse concept is illustrated in Fig. 1.2. In this figure, cells shown with the same letter have the same set of frequencies. The minimum grouping of adjacent cells which collectively use the complete available spectrum is called a *cluster*, the cluster size N being the number of cells it contains. Here the cluster size is three, and three clusters are shown. Since there are three sets of frequencies, each cell uses 1/3 of total number of available frequencies, and the *frequency reuse factor* is 1/3.

In frequency reuse, the cells that use the same set of frequencies are called *co-channel cells*, and the interference caused in one such cell by other such cells is called *co-channel interference* (CCI). If, in a given frequency reuse system, the size of each cell is approximately identical and the base stations transmit equal power, the signal to co-channel interference ratio becomes solely a function of the cell radius (R) and the distance between the centers of the nearest co-channel cells (D). Interference is reduced by increasing the ratio D/R and vice versa. This ratio D/R is

Fig. 1.2 Frequency
reuse concept

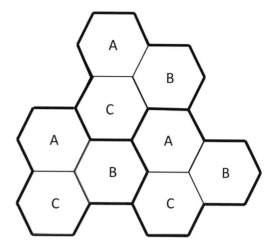

referred to as the *co-channel reuse ratio*, *Q*. Decreasing the value of Q implies
decreasing cluster size, resulting in more reuse of frequencies in a given large area
and hence more capacity. This capacity increase is achieved, however, at the expense
of increased interference. We note that, in a multi-cluster layout, a cluster size of 3
is the smallest size that affords no co-channel interference from an adjacent cell.

1.5.2   Coverage Enhancements

As the demand for greater and greater capacity grows, the traditional cellular
approach of simply increasing cell density often proves insufficient and or unafford-
able to meet the need. As a result, several innovate approaches have been devised to
address this issue. Some of these approaches are outlined below.

1.5.2.1   5G Frequency Reuse Schemes

Users within a cell, in both UL and DL, are "orthogonal" to each other as they all
communicate, as we shall see, with the BS at any one time on different frequencies.
Thus, there is little or no interference between transmissions within a cell, that is,
there is little or no *intra-cell interference*. Mobile system performance is limited,
however, in terms of spectral efficiency and available data rates, by intercell co-
channel interference.

5G systems are designed to operate with a much higher degree of intercell co-
channel interference than preceding generations. This has led to the use of frequency
reuse strategies that result in reuse factors approaching 1.

A simple and effective scheme to reduce intercell co-channel interference at the
cell edges in 4G and 5G systems is to employ the traditional frequency reuse

approach, with a cluster size of 3 and reuse factor of 1/3 as shown in Fig. 1.2. However, because each base station uses only one-third of the total available bandwidth, the capacity and peak data rate (as opposed to average user data rate) of the covered cell are each reduced by a factor of three.

A second and much more extreme approach is to simply use a system with a reuse factor of 1. Here, all cells use the full frequency allotment. This approach allows the system to offer maximum capacity and maximum peak data rate, but due to heavy interference at the cell edges from adjacent cells, cell-edge user performance is greatly limited.

One scheme to reduce the heavy cell-edge co-channel interference while keeping the reuse factor close to 1 is called *fractional frequency reuse* (FFR) [2] or sometimes *Strict FFR*. This scheme, which is illustrated in Fig. 1.3, is a combination of the two schemes outlined immediately above. Here, each cell is conceptually divided into two parts, the cell interior and the cell edge. The system bandwidth is also divided into sub-bands. One band, typically two-thirds of the total band, is used only in the cell interiors of all cells. The other one-third of the band is divided into three sub-bands, and each provides coverage at the cell edge of one of the three cells making up a cluster as shown in Fig. 1.3. Relative to the power transmitted to the cell edges, that transmitted to cell interior users is reduced since such users are closer to the base station and lower power reduces interference to interior users in adjacent cells.

A variation of FFR which results in a somewhat higher frequency reuse is *soft frequency reuse* (SFR) [2]. SFR, which is illustrated in Fig. 1.4, is identical to FFR with the exception that the interior users are allowed to use the entire frequency band. Though more bandwidth efficient than FFR, SFR results in more interference to both edge users and interior users.

An alternative and often utilized layout model applicable to FFR and SFR is to locate three sectored antennas at the junction of three cells, each antenna radiating one of the cells. This structure is referred to as the clover leaf model and is illustrated in Fig. 1.5 for FFR.

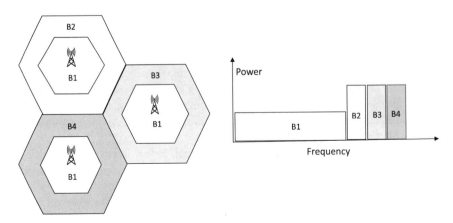

Fig. 1.3 Fractional frequency reuse

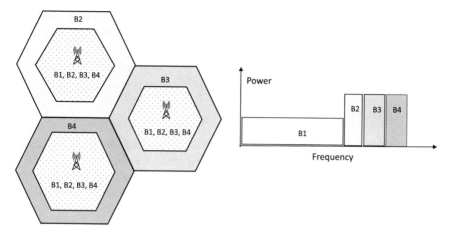

Fig. 1.4 Soft frequency reuse

Fig. 1.5 FFR with clover leaf model

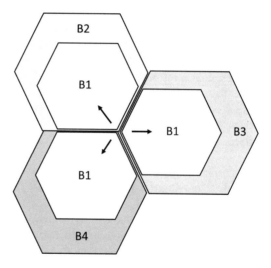

A number of other frequency reuse schemes, somewhat similar to the two outlined above, have been proposed, but all share the same objective, namely that of increasing the reuse factor to as close to 1 as possible while keeping cell-edge co-channel interference to a tolerable level.

1.5.2.2 Relaying

Relaying is the process whereby relay nodes are used to extend the coverage area of a limited segment of a cell. The nodes transfer signals from a donor BS to the UE and from the UE to the donor BS with the connection between the BS and the relay node being a wireless one. From the UEs point of view, the relay node is normally made to appear as just another BS. The relay node thus functions as a BS, normally of lowered power, connected wirelessly to the donor BS. Relay nodes find application is situations where performance is primarily limited by coverage, not capacity. Figure 1.6 illustrates a typical relay scenario.

1.5.2.3 Heterogeneous Deployment

The traditional cellular concept way of increasing system capacity in a given area is by increasing the number of macrocells in that area. A network so structured is referred to as a *homogeneous* or *single-layer* one. In scenarios where some users are highly clustered, an attractive alternative is to complement the higher transmit power macrocells with lower transmit power cells such as so called *picocells* where needed. Networks so structured are referred to as *heterogeneous* or *multi-layered* ones. Figure 1.7 illustrates homogeneous and heterogeneous deployment.

If the available frequency resources are divided and used for different layers, interlayer interference can be avoided. This, however, could leads to a reduction in the peak data rate per layer. It is therefore preferable to use of the same frequency resources in all layers. The obvious disadvantage to this approach, however, is interference between layers. In this situation, the signal-to-interference ratio experienced by a terminal at the outermost coverage area of a low-power cell will be measurably lower than that in a standard homogeneous network, due to the difference in power output between high transmit power macro nodes and low transmit power pico nodes. Special techniques are thus usually applied to minimize these interference effects.

1.5.2.4 Coordinated Multipoint (CoMP) Transmission/Reception

Coordinated multipoint (CoMP) transmission/reception, applicable to both the DL and UL, is a tool to improve coverage, cell-edge throughput, and minimize intercell interference. As the name implies, CoMP seeks to optimization the transmission and reception from multiple transmission points (TPs) in a coordinated fashion.

Fig. 1.6 Typical relay scenario

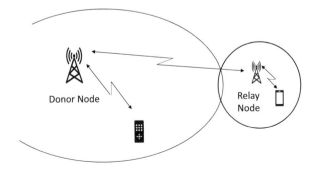

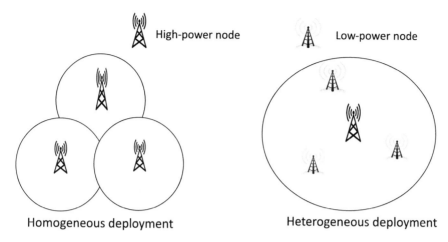

Fig. 1.7 Homogeneous and heterogeneous deployment

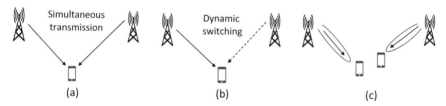

Fig. 1.8 DL CoMP: (**a**) joint transmission, (**b**) dynamic point selection, (**c**) coordinated scheduling/beamforming

Deployment scenarios include configurations where TPs are high-power cells, as well as heterogeneous configurations, where TPs can be both high power and low power.

In coordinated DL transmission, signals transmitted from multiple TPs are coordinated to improve at the UE the signal strength of the desired signal and/or reduce co-channel interference. In 3GPP networks there are three schemes for DL CoMP, namely, *joint transmission* (JT), *dynamic point selection* (DPS), and *coordinated scheduling/beamforming* (CS/CB). Figure 1.8 depicts these schemes.

With joint transmission, user data is transmitted from multiple TPs simultaneously and coherently to a single UE. Since transmission is in the same time-frequency resource from multiple TPs, this approach is best employed in lightly loaded networks.

With dynamic point selection, the user data is available at multiple TPs but, to reduce interference, is only transmitted to a single UE from one TP at a time. Transmission can, however, be switched dynamically between the multiple TPs.

With coordinated scheduling/beamforming, like DPS, user data is only transmitted to a single UE from one TP at a time. However, scheduling, including any beamforming, is dynamically coordinated between the cooperating cells to control and/or

Fig. 1.9 UL CoMP

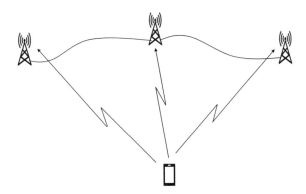

reduce interference between the different transmitted signals. Transmitted beams are constructed with the aim of reducing interference to other nearby users while increasing the signal strength to the individual served user.

In coordinated UL transmission, the signal transmitted from a single UE is processed by multiple reception points so as to improve the reliability of the received signal. Techniques applied fall into two categories: dynamic point selection, where the best received signal is chosen based on short-term channel quality, or joint reception, where signals received at the several reception points are combined for optimum signal quality. Figure 1.9 depicts UL CoMP.

1.6 Summary

In this chapter mobile and fixed wireless access networks were introduced. It provided a top-level description of such networks and a brief history of their evolution up to the era of 5G, indicating at a high level the significant performance improvement offered by 5G. Finally, several key cellular systems coverage methods relevant to 5G deployment were reviewed to help establish the environment that mobile and fixed wireless networks operate in.

References

1. Rappaport TS (2002) Wireless communications; principles and practice. Prentice Hall PTR, Upper Saddle River
2. Chandra T, Chandrasekhar C (2012) Comparative evaluation of fractional frequency reuse (FFR) and traditional frequency reuse in 3GPP-LTE downlink. Int J Mobile Netw Commun Telematics (IJMNCT) 2(4):45–52

Chapter 2
Broadband Wireless Payload: Packet-Switched Data

2.1 Introduction

Traditional telephone communication is based on *circuit switching*, where circuits are set up and kept continually open between users until the parties end their communications and release the connection. Traditional data communication, however, is based on *packet switching* [1, 2], where the data is sent in discrete packets, each packet being a sequence of *bytes* (each byte contains 8 bits). Contained within each packet is its sender's address and the address of the intended recipient, and each packet seeks the best available route to its destination. In packet-switched networks, the packets at the receiving end don't necessarily arrive in the order sent, so they must be reordered before further use. Here, unlike circuit switching, a circuit is configured and left open only long enough to transport an addressed packet and then freed up to allow reconfiguration for transporting the next packet, which may be from a different source and addressed to a different destination. Packet data communication systems employ so-called protocols. Here a protocol is a set of rules that must be followed for two devices to communicate with each other. Protocol rules are implemented in software and firmware and cover (a) data format and coding, (b) control information and error handling, and (c) speed matching and sequencing. Protocols are layered on top of each other to form a communication architecture referred to as a *protocol stack*. Each protocol provides a function or functions required to make data communication possible. Typically, many protocols are used so that overall functioning can be broken down into manageable pieces.

5G networks communicate via the Internet. The Internet communicates with the aid of a suite of software network protocols called *Transmission Control Protocol/Internet Protocol*, or *TCP/IP*. With TCP/IP, data to be transferred along with administrative information is structured into a sequence of so-called *datagrams*, each

---

The original version of this chapter was revised. The correction to this chapter is available at https://doi.org/10.1007/978-3-030-51441-9_12

© Springer Nature Switzerland AG 2020, Corrected Publication 2021
D. H. Morais, *Key 5G Physical Layer Technologies*,
https://doi.org/10.1007/978-3-030-51441-9_2

datagram being a sequence of bytes. TCP/IP works together with a physical data link layer to create an effective packet-switched network for Internet communications. The data link layer protocol puts the TCP/IP datagrams into packets (it's like putting an envelope into another envelope) and is responsible for the reliable transmission of packets over the physical layer. The data link protocol associated with TCP/IP in the 5G world is the *Ethernet* protocol. In the following sub-sections, a brief introduction to TCP/IP, including the Ethernet protocol, is presented.

2.2 TCP/IP

The acronym TCP/IP is commonly used in the broad sense and as applied here refers to a hierarchy of four protocols but derives its name from the two main ones, namely, TCP and IP. The four protocols are stacked as shown in Fig. 2.1. Following is a short description of each protocol layer:

2.2.1 Application Layer Protocol

This is where the user interfaces with the network and includes all the processes that involve user interaction. Protocols at this level include those for facilitating World Wide Web (WWW) access, e-mail, etc. Data from this level to be sent to a remote address is passed on the next lower layer, the transport layer, in the form of a stream of 8-bit bytes.

2.2.2 Transport Layer Transmission Control Protocol

As seen in Fig. 2.1, the *Transmission Control Protocol* (TCP) is one of two protocols at this layer. TCP provides reliable end-to-end communication and is used by most Internet applications. With TCP, two users at each end of the network must establish a two-way connection before any data can be transferred between them. It is thus referred to as a *connection-oriented* protocol. It guarantees that data sent is not only received at the far end but correctly so. It does this by having the far end acknowledge receipt of the data. If the sender receives no acknowledgment within a specified time frame, it resends the data. When TCP receives data from the application layer above for transmission to a remote address, it first splits the data into manageable blocks based on its knowledge of how large a block the network can handle. It then adds control information to the front of each data block. This control information is called a *header*, and the addition of the header to the data block is called *encapsulation*. TCP adds a 20-byte header to the front of each data block to form a TCP datagram. It then passes these datagrams to the next layer below, the IP

Fig. 2.1 TCP/IP protocol stack

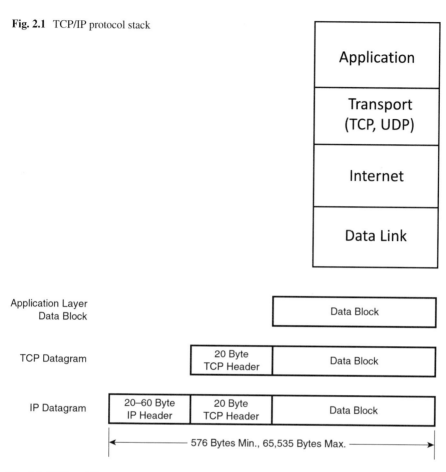

Fig. 2.2 TCP and IP encapsulation

layer. A TCP datagram is shown Fig. 2.2. When TCP receives a datagram from a remote address via the IP layer below, the opposite procedure to that involved in creating a datagram takes place, i.e., the header is removed and the data passed to the application layer above. However, before passing it above, TCP takes data that arrives out of sequence and puts it back into the order in which it was sent.

2.2.3   Transport Layer User Datagram Protocol

Unlike TCP, with the *User Datagram Protocol* (UDP), users at each end of the network need not establish a connection before any data can be transferred between them. It is therefore described as a *connectionless* protocol. It provides unreliable service with no guarantee of delivery. Packets may be dropped, delayed, or may arrive out of sequence. The service it provides is referred to as a best effort one as all attempts are made to deliver packets, with poor reliability caused only by hardware faults or exhausted resources and the fact that no retransmission is requested in the event of error or loss. UDP adds an 8-byte header to the front of each data block to form a UPD datagram. UPD packets arrive more quickly than TCP ones and are processed faster at the cost of potential error. UPD is used primarily for connections where low latency is important, and some data loss is tolerable. Good examples of such connections are voice and video.

2.2.4 Internet Layer Protocol

The protocol at this layer, the *Internet Protocol* (IP), is the core protocol of TCP/IP. Its job is to route data across and between networks. When sending datagrams, it figures out how to get them to their destination; when receiving datagrams, it figures out to whom they belong. It is an unreliable protocol, unconcerned as to whether datagrams it sent arrive at their destination and whether datagrams it received arrive in the order sent. If a datagram arrives with any problems, it simply discards it. It leaves the quest for reliable communication, if required, to the TCP level above. Like UDP, it is defined as a *connectionless* protocol. There are, naturally, provisions to create connections per se or communication would be impossible. However, such connections are established on a datagram by datagram basis, with no relationship to each other. It processes each datagram as an entity, independent of any other datagram that may have preceded it or may follow it. In fact, there is no information in an IP datagram to identify it as part of a sequence or as belonging to a particular task. Thus, for IP to accomplish its assigned task, each IP datagram must contain complete addressing information. An IP datagram is created by adding a header to the transport layer datagram received from above.

Internet Protocols are developed by the *Internet Engineering Task Force* (IETF). *IPv4* was the fourth iteration of IP developed by the IETF but the first version to be widely deployed. It is the dominant network layer protocol on the Internet. Its header is normally 20 bytes long but can in rare circumstances be as large as 60 bytes. A later IP version is IPv6. Its header is 40 bytes long, but "extension" headers can be added when needed.

An IPv4 datagram created from a TCP datagram is shown in Fig. 2.2. This header includes source and destination address information as well as other control information. The maximum size of a TCP-derived IP datagram is 65,535 bytes. All IP networks must be able to handle such IP datagrams of at least 576 bytes in length. IP passes datagrams destined to a remote address to the next layer below, the link layer. On the receiving end, datagrams received by the IP layer from the link layer are stripped of their IP headers and passed up to the transport layer. The maximum amount of data that a link layer packet can carry is called the *maximum transfer unit* (MTU) of the layer. Because, as is explained below, each IP datagram is encapsulated within a link layer packet prior to transmission to the remote address, the MTU of the link layer places a maximum on the length of an IP datagram that it can process. Should an IP datagram be larger than this maximum, it is broken up into smaller datagrams that can fit in the link layer packet. This breaking up process is called *fragmentation*, and each of the smaller datagrams created is called a *fragment*. Figure 2.3 shows the fragmentation of an IP datagram 1000 bytes long into two fragments, in order that it may be processed by a link layer with an MTU of 576 bytes. It will be noticed that each fragment has the same basic structure as an IP datagram. Information in the fragment header defines it as a fragment, not a datagram. At the destination, fragments are reassembled to the original IP datagram before being passed to the transport layer. Should one or more of the fragments fail to arrive, the datagram is assumed to be lost, and nothing is passed to the transport layer. Fragmentation puts additional tasks on Internet hardware, and so it is desirable to keep it to a minimum. Obviously, it can be entirely eliminated by using IP datagrams no longer than 576 bytes, since all IP handling networks must have an MTU of at least 576. In fact, most bulk TCP/IP data transfer is done using IP datagrams 576 bytes long.

For an IP datagram created from a UDP datagram, the data block size can be as small as zero bytes. As the UDP header is only 8 bytes long and the IPv4 header can be as little as 20 bytes, a UDP-generated IPv4 datagram can be as small as 28 bytes.

IPv4 uses 32-bit addresses which limits address space to 4295 million possible unique addresses. As addresses available are consumed, it appears that an IPv4 address shortage is inevitable. Further, IPv4 provides only limited quality of service (QoS) capability. It is the limitation in address space and QoS capability of IPv4 that helped stimulate the push to *IPv6* which is the only other standard internet network layer used on the Internet. IPv6 uses 128-bit addresses which results in approximately 300 billion, billion, billion, billion unique addresses! Further, IPv6 provides for true QoS. IPv4 will be supported alongside IPv6 for the foreseeable future.

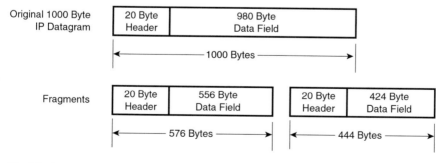

Fig. 2.3 An example of the fragmentation of an IPv4 datagram

An IPv4-to-IPv6 transition technology that involves the presence of IPv4 and IPv6 operating in parallel in an operating system is called *dual-stack IPv4/IPv6*. It enables the application to choose which of the two IP protocols to use or automatically selects it according to address type.

2.2.5 Data Link Layer Ethernet Protocol

The data link layer consists of the network hardware that effects actual communication. It stipulates the details of how data is physically sent over the network, including how bits are electrically or optically handled by devices that interface directly with a network link such as twisted pair copper wire, coaxial cable, or optical fiber. When transmitting IP data onto the network, it creates packets by taking IP datagrams and adding headers and, in some instances, trailers (control data trailing the datagrams) on to them and dispatches these packets out onto the network. When data packets are received from the network, it strips them of their headers and trailers, if any, and passes them on to the IP layer above. As indicated above, the data link layer protocol associated with TCP/IP in the 5G world is the Ethernet protocol.

Ethernet packets range in size from 72 to 1526 bytes. Figure 2.4 shows the Ethernet packet format. It consists of a 22-byte header, followed by information data, and ending in a 4-byte trailer. The header commences with a 7-byte preamble used for synchronization of the receiving station's clock, followed by a 1-byte start frame delimiter used to indicate the start of the frame. This is followed by two 6-byte address codes, the first representing the packet's destination and the second representing the packet's source. Next is a 2-byte length field used to indicate the length of the information data field. Following this is the information data field, which is an integer number of bytes, from a minimum of 46 to a maximum of 1500. Note that if the actual number of data bytes to be sent is less than 46, then extra bytes are added at the end to total 46. The 4-byte trailer follows the information data field and forms the frame check sequence, which provides the error detection and

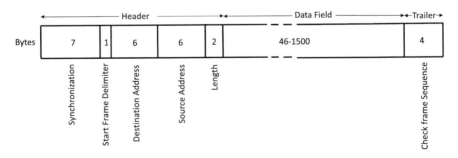

Fig. 2.4 Ethernet packet format

correction at the bit level to ensure that the data from the destination address through the information data field reaches its destination correctly.

Ethernet communication speed ranges all the way from 10 Mb/s to several Gb/s. With 4G networks the Ethernet speed typically employed varies from about 100 Mb/s up to 10 Gb/s. Connections at 100 and 1000 Mb/s are electrical or optical, but 10 GHz connections are optical only. 5G networks are likely to use optically connected Ethernet only with speeds between 10 Gb/s and 100 Gb/s.

2.3   Voice over IP (VoIP)

In the narrowest sense, *Voice over Internet Protocol* (VoIP) implies voice communication over only an IP link. In the broader and more realistic sense, it means voice communication where at least a part of the network is an IP link. Conceptually, communication over an all IP VoIP network is quite straightforward:

- The analog voice signal is converted to a digital stream, typically at 64 kb/s.
- This data stream is then typically compressed using a *codec* (a codec is a compression/decompression device).
- Bits in this compressed stream are grouped into packets.
- These packets are transported via an IP network to destination.
- At destination they are converted back into the original data stream and finally, back into an analog signal.

The structure of an IPv4-based VoIP packet is shown in Fig. 2.5. Note that immediately ahead of the voice payload data, a new header has been added. This header is called the *Real-time Transport Protocol* (RTP) [3] and was developed by the IETF. It's typically 12 bytes long. RTP is a protocol to facilitate the delivery of audio and video over IP networks and typically runs over UDP. It addresses jitter compensation, detection of packet loss, and out of order delivery, issues which are likely with UDP transmissions. We note that the total packet header size is 40 bytes.

As 5G networks only support packet services, voice service provided over these networks must be via VoIP. Adaptive Multirate (AMR) codecs are the ones most commonly used for creating VoIP data. AMR codecs come in two versions, namely, AMR-Narrowband (AMR-NB) [4] and AMR-Wideband (AMR-WB) [5].

IP Header (20 bytes)	UDP Header (8 bytes)	RTP Header (12 bytes min.)	Real Time Payload e.g. voice, video

Fig. 2.5 Structure of an IPv4 voice packet

AMR-NB uses a sampling rate of 8 kHz and provides an audio bandwidth of 300–3400 Hz. Its output data rates range from 4.75 to 12.2 kb/s. AMR-WB uses a sampling rate of 16 kHz and provides an audio bandwidth of 50–7000 Hz. Its output data rates range from 6.6 to 28.85 kb/s, with a commonly used rate of 12.65 kb/s. For comparable data rates, AMR-WB offers substantially better voice quality than AMR-NB, which is not surprising, given AMR-WB's much wider audio bandwidth.

The size of the voice payload data sent over VoIP packets depends on the coding rate and the voice sampling period. Typically, the voice sampling period is 20 ms. With a coding rate of 12.65 kb/s, 253 bits are generated every 20 ms. This computes to 31.625 bytes, but as data payload is transmitted in an integer number of bytes, this would be transmitted as 32 bytes. Thus, with IPv4, and therefore a packet header size of 40 bytes, the total VoIP packet would be 72 bytes, and the header overhead would be 55% of the total packet. Had we assumed IPv6, then the packet header size would have been 60 bytes leading to a header overhead of 65%. Because this high overhead is clearly wasteful of capacity, header compression is normally used along with VoIP. This is normally accomplished via *robust header compression* (ROHC) [6], a compression technique used in streaming applications such as voice and video. It compresses 40 or 60 bytes of overhead typically into only 1 or 3 bytes!

2.4   Video over IP

Video over IP from a technical point of view is identical to voice over IP with the exception that we are dealing with much higher data rates. These higher rates are the result of the inherently much higher information content of video compared to voice. Uncompressed video rates are very high, and hence much compression is required to achieve manageable transmission rates. Video signal compression is a complex process and will not be addressed here. Suffice it to say, however, that it is possible because there are redundancies in a video signal's uncompressed format and because visual perception has certain characteristics that can be used to advantage. Regarding format redundancies, video compression in essence results in the transmission of only differences from one video frame to the next, rather than the transmission of each new frame in its entirety.

Many video codec standards are available, but to get a sense of their capability, we will look at the compression capability of one of the most popular, namely, the *H.264 Advanced Video Coding* (H.264/AVC) standard. To do this we will first determine uncompressed video data rate, VR_{UC} say. VR_{UC} is given by:

$$VR_{UC} = \text{color depth} \times \text{number of vertical pixels}$$
$$\times \text{number of horizontal pixels} \times \text{refresh rate (frames per second)} \quad (2.1)$$

where:

• Color depth is the number of bits used to indicate the color of a single pixel.

- Refresh rate is the rate at which the screen presentation is refreshed.

For the 1080p high definition (HD) video format, with color depth of 24, 1080 vertical pixels, 1920 horizontal pixels, and a refresh rate of 30, VR_{UC} computes to 1493 Mb/s. Video compression possible with H.264 varies. H.246 offers a number of profiles. Its "baseline" profile, for example, can achieve a compression rate of up to about 1000:1. Its "high" profile can achieve a compression rate of up to about 2000:1. These are impressive numbers, but as compression rate increases, picture quality decreases, so there is a practical limit to how much compression is normally used. For the 1080p format outlined above, a compressed bit rate of about 5 Mb/s or more is required for minimum acceptable picture quality. Assuming a rate of 5 Mb/s, then the compression rate is 1493/5, or approximately 300. For 4 K ultra HD (2160 vertical x 3890 horizontal pixels), a compressed bit rate of about 15 Mb/s is required for minimum acceptable quality, and a rate of about 25 Mb/s required for good quality.

2.5 Summary

5G networks communicate via the Internet with the aid of the TCP/IP and the Ethernet protocols. In this chapter these protocols were reviewed. Further, since voice- and video-generated data are some of the more commonly transported via the Internet to and from 5G networks, how this data is generated was also reviewed.

References

1. Lee BG, Kang M, Lee J (1993) Broadband telecommunications technology. Artech House, Inc, Norwood
2. Bates B, Gregory D (1996) Voice & data communications handbook. Mc-Graw-Hill, Inc, New York
3. Schulzrinne H et al. (2003) IETF RC 3550 RTP: A Transport Protocol for Real-Time Applications, July 2003
4. 3GPP TS 26.071 (2018) Mandatory speech CODEC speech processing functions; AMR speech CODEC; General description Version 1500, 22 June 2018
5. 3GPP TS 26.190 (2018) Speech codec speech processing functions; Adaptive Multi-Rate – Wideband (AMR-WB) speech codec; Transcoding functions Version 1500, 22 June 2018
6. IETF RFC 5795 (2010) The Robust Header Compression (ROHC) Framework. Mar 2010

Chapter 3
Mathematical Tools for Digital Transmission Analysis

3.1 Introduction

The study of digital wireless transmission is in large measure the study of (a) the conversion in a transmitter of a binary digital signal (often referred to as a *baseband signal*) to a modulated RF signal, (b) the transmission of this modulated signal from the transmitter through the atmosphere, (c) the corruption of this signal by noise and other unwanted signals, (d) the reception of this corrupted signal by a receiver, and (e) the recovery in the receiver, as best as possible, of the original baseband signal. In order to analyze such transmission, it is necessary to be able to characterize mathematically in the time, frequency and probability domains, baseband signals, modulated RF signals, noise, and signals corrupted by noise. The purpose of this chapter is to briefly review the more prominent of those analytical tools used in such characterization, namely, spectral analysis and relevant statistical methods.

Spectral analysis permits the characterization of signals in the frequency domain and provides the relationship between frequency domain and time domain characterizations. Noise and propagation anomalies are random processes leading to uncertainty in the integrity of a recovered signal. Thus, no definitive determination of the recovered signal can be made. By employing statistical methods, however, computation of the fidelity of the recovered baseband signal is possible in terms of the probability that it's in error. The study of non-light-of-sight propagation analysis set out in Chap. 4, the basic principles of wireless modulation covered in Chap. 5, and the multiple access techniques outlined in Chap. 7 will apply several of the tools presented here. Those readers familiar with these tools may want to skip this chapter and proceed to Chap. 4.

© Springer Nature Switzerland AG 2020
D. H. Morais, *Key 5G Physical Layer Technologies*,
https://doi.org/10.1007/978-3-030-51441-9_3

3.2 Spectral Analysis of Nonperiodic Functions

A nonperiodic function of time is a function that is nonrepetitive over time. A stream of binary data as typically transmitted by digital communication systems is a stream of nonperiodic functions, each pulse having equal probability of being one or zero, independent of the value of other pulses in the stream. The analysis of the spectral properties of nonperiodic functions is thus an important component of the study of digital transmission.

3.2.1 The Fourier Transform

A nonperiodic waveform, *v(t)* say, may be represented in terms of its frequency characteristics by the following relationship

$$v(t) = \int_{-\infty}^{\infty} V(f) e^{j2\pi ft} df \tag{3.1}$$

The factor *V(f)* is the *amplitude spectral density* or the *Fourier transform* [1] of *v(t)*. It is given by

$$V(f) = \int_{-\infty}^{\infty} v(t) e^{-j2\pi ft} dt \tag{3.2}$$

Because *V(f)* extends from $-\infty$ to $+\infty$, i.e., it exists on both sides of the zero frequency axis, it is referred to as a *two-sided* spectrum.

An example of the application of the Fourier transform that is useful in the study of digital communications is its use in determining the spectrum of a nonperiodic pulse. Consider a pulse *v(t)* shown in Fig. 3.1a, of amplitude *V*, and that extends from $t = -\tau/2$ to $t = \tau/2$. Its Fourier transform, *V(f)*, is given by

$$V(f) = \int_{-\tau/2}^{\tau/2} V e^{-j2\pi ft} dt$$

$$= \frac{V}{-j2\pi f} \left[e^{-j2\pi f\tau/2} - e^{j2\pi f\tau/2} \right] \tag{3.3}$$

$$= V\tau \frac{\sin \pi f\tau}{\pi f\tau}$$

The form (sin *x*)/*x* is well-known and referred to as the *sampling function, Sa(x)* [1]. The plot of *V(f)* is shown in Fig. 3.1b. It will be observed that it is a continuous function. This is a common feature of the spectrum of all nonperiodic waveforms. We note also that it has zero crossings at $\pm 1/\tau, \pm 2/\tau, \ldots$.

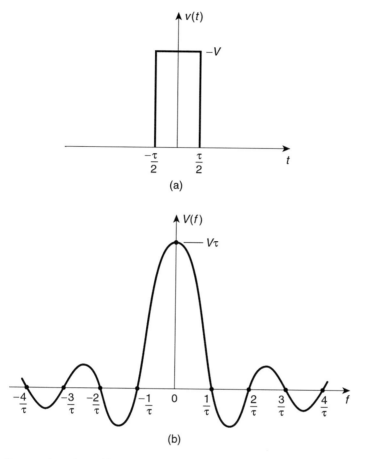

Fig. 3.1 Rectangular pulse and its spectrum. (**a**) Rectangular pulse. (**b**) Spectrum of rectangular pulse in (**a**)

The Fourier transform $V(f)$ of an impulse of unit strength is also a useful result. By definition, an impulse $\delta(t)$ has zero value except at time $t = 0$, and an impulse of unit strength has the property

$$\int_{-\infty}^{\infty} \delta(t) dt = 1 \tag{3.4}$$

Thus

$$V(f) = \int_{-\infty}^{\infty} \delta(t) e^{-2\pi jft} dt = 1 \tag{3.5}$$

Equation (3.5) indicates that the spectrum of an impulse $\delta(t)$ has a constant amplitude and phase and extends from $-\infty$ to $+\infty$.

A final example of the use of the Fourier transform is the analysis of what results in the frequency domain when a signal $m(t)$, with Fourier transform $M(f)$, is multiplied by a sinusoidal signal of frequency f_c. In the time domain the resulting signal is given by

$$v(t) = m(t).\cos 2\pi f_c t$$
$$= m(t)\left[\frac{e^{j2\pi f_c t} + e^{-j2\pi f_c t}}{2}\right] \tag{3.6}$$

and its Fourier transform is thus

$$V(f) = \frac{1}{2}\int_{-\infty}^{\infty} m(t)e^{-j2\pi(f+f_c)t}dt + \frac{1}{2}\int m(t)e^{-j2\pi(f-f_c)t}dt \tag{3.7}$$

Recognizing that

$$M(f) = \int_{-\infty}^{\infty} m(t)e^{-j2\pi ft}dt \tag{3.8}$$

then

$$V(f) = \frac{1}{2}M(f+f_c) + \frac{1}{2}M(f-f_c) \tag{3.9}$$

An amplitude spectrum $|M(f)|$, band limited to the range $-f_m$ to $+f_m$, is shown in Fig. 3.2a. In Fig. 3.2b, the corresponding amplitude spectrum of $|V(f)|$ is shown.

3.2.2   The Discrete and Inverse Discrete Fourier Transform

The *Discrete Fourier Transform* (DFT), like the Fourier transform, transforms a signal from the time domain to the frequency domain. However, as implied by its nomenclature, it requires an input function that is *discrete* and whose non-zero values have a limited, i.e., finite, duration and creates an output function that's also discrete. Since its input function is a finite series of real or complex numbers, the DFT is widely employed in signal processing to analyze the frequencies contained in a sampled signal. The time domain sequence N of complex numbers $x_0,\ldots, x_{N-1}$ is transformed into the frequency domain sequence of N complex numbers $X_0,\ldots, X_{N-1}$ by the DFT according to the formula:

$$X_k = \sum_{n=0}^{N-1} x_n e^{-j\frac{2\pi kn}{N}} \quad k = 0,\ldots, N-1 \tag{3.10}$$

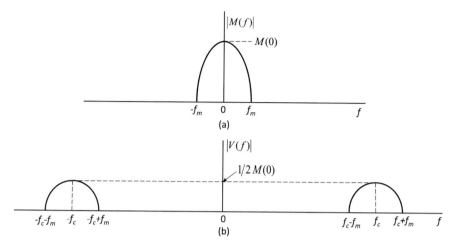

Fig. 3.2 (**a**) The amplitude spectrum of a waveform with no special component beyond f_m (**b**) The amplitude spectrum of the waveform in (**a**) multiplied by $\cos 2\pi f_c$

The *Inverse Discrete Fourier transform* (IDFT) transforms a signal from the frequency domain to the time domain and is given by:

$$x_n = \frac{1}{N}\sum_{k=0}^{N-1} X_k e^{j\frac{2\pi kn}{N}} \quad n = 0,\ldots,N-1 \tag{3.11}$$

A simple description of the above equations is that the complex numbers X_k represent the amplitude and phase of different sinusoidal components of the input signal x_k. The DFT computes X_k from x_n, a finite series of time domain samples, while the IDFT computes x_n from X_k, a finite series of sinusoidal components. Standard digital signal processing (DSP) notation uses lowercase letters to represent time domain signals such as $x[\ ]$ and $y[\ ]$. Corresponding uppercase letters, i.e., $X[\ ]$ and $Y[\ ]$, are used to represent their frequency domain components. The DFT/IDFT can be computed efficiently in practice by using the Fast Fourier Transform (FFT)/Inverse Fast Fourier Transform (IFFT).

3.2.3   Linear System Response

A *linear system* is one in which, in the frequency domain, the output amplitude at a given frequency bears a fixed ratio to the input amplitude at that frequency and the output phase at that frequency bears a fixed difference to the input phase at that frequency, irrespective of the absolute value of the input signal. Such a system can be characterized by the complex transfer function, $H(f)$ say, given by

Fig. 3.3 Signal transfer
through a linear system

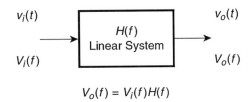

$$V_o(f) = V_i(f)H(f)$$

$$H(f) = \left| H(f) \right| e^{-j\theta(2\pi f)} \tag{3.12}$$

where |H(f)| represents the absolute amplitude characteristic and $\theta(2\pi f)$ the phase characteristic of H(f).

Consider a linear system with complex transfer function H(f), as shown in Fig. 3.3, with an input signal $v_i(t)$, an output signal $v_o(t)$, and with corresponding spectral amplitude densities of $V_i(f)$ and $V_o(f)$. After transfer through the system, the spectral amplitude density of $V_i(f)$ will be changed to $V_i(f)H(f)$. Thus

$$V_o(f) = V_i(f)H(f) \tag{3.13}$$

and

$$v_o(t) = \int\limits_{-\infty}^{\infty} V_i(f) H(f) e^{j2\pi ft} df \tag{3.14}$$

An informative situation is the one where the input to a linear system is an impulse function of unit strength. For this case, as per Eq. (3.5), $V_i(f) = 1$, and

$$V_o(f) = H(f) \tag{3.15}$$

Thus, the output response of a linear system to a unit strength impulse function is the transfer function of the system.

3.2.4   Energy and Power Analysis

In considering energy and power in communication systems, it is often convenient to assume that the energy is dissipated in a 1-ohm resistor; as with this assumption, one need not keep track of the impact of the true resistance value, R say. When this assumption is made, we refer to the energy as the *normalized energy* and to the power as *normalized power*. It can be shown that the normalized energy E of a non-periodic waveform v(t), with a Fourier transform V(f), is given by

$$E = \int\limits_{-\infty}^{\infty} \left[v(t) \right]^2 dt = \int\limits_{-\infty}^{\infty} \left| V(f) \right|^2 df \tag{3.16}$$

The above relationship is called *Parseval's theorem* [1]. Should the actual energy be required, then it is simply E (as given in Eq. 3.16 above) divided by R.

The *energy density*, $D_e(f)$, of a waveform is the factor $dE(f)/df$. Thus, by differentiating the right-hand side of Eq. (3.16), we have

$$D_e(f) = \frac{dE(f)}{df} = |V(f)|^2 \tag{3.17}$$

For a nonperiodic function such as a single pulse, normalized energy is finite, but power, which is energy per unit time, approaches zero. Power is thus somewhat meaningless in this context. However, a train of binary nonperiodic adjacent pulses does have meaningful average normalized power. This power, P say, is equal to the normalized energy per pulse E, multiplied by f_s, the number of pulses per second, i.e.,

$$P = Ef_s \tag{3.18}$$

If the duration of each pulse is τ, then $f_s = 1/\tau$. Substituting this relationship and Eq. (3.16) into Eq. (3.18), we get

$$P = \frac{1}{\tau} \int_{-\infty}^{\infty} |V(f)|^2 \, df \tag{3.19}$$

The *power spectral density* (PSD), $G(f)$, of a waveform is the factor $dP(f)/df$. Thus, by differentiating the right-hand side of Eq. (3.19), we have

$$G(f) = \frac{dP(f)}{df} = \frac{1}{\tau} |V(f)|^2 \tag{3.20}$$

To determine the effect of a linear transfer function $H(f)$ on normalized power, we substitute Eq. (3.13) into Eq. (3.19). From this substitution, we determine that the normalized power, P_o, at the output of a linear network, is given by

$$P_o = \frac{1}{\tau} \int_{-\infty}^{\infty} |H(f)|^2 \, |V_i(f)|^2 \, df \tag{3.21}$$

Also, from Eq. (3.13), we have

$$\frac{|V_o(f)|^2}{\tau} = \frac{|V_i(f)|^2}{\tau} |H(f)|^2 \tag{3.22}$$

Substituting Eq. (3.20) into Eq. (3.22), we determine that the power spectral density $G_o(f)$ at the output of a linear network is related to the power spectral density $G_i(f)$ at the input by the relationship

$$G_o(f) = G_i(f)|H(f)|^2 \tag{3.23}$$

3.3   Statistical Methods

We now turn our attention away from the time and frequency domain and towards the probability domain where statistical methods of analysis are employed. As indicated in Sect. 3.1 above, such methods are required because of the uncertainty resulting from the introduction of noise and other factors during transmission.

3.3.1   *The Cumulative Distribution Function and the Probability Density Function*

A random variable X [1, 2] is a function that associates a unique numerical value $X(\lambda_i)$ with every outcome λ_i of an event that produces random results. The value of a random variable will vary from event to event and, depending on the nature of the event, will be either *discrete* or *continuous*. An example of a discrete random variable X_d is the number of heads that occur when a coin is tossed four times. As X_d can only have the values 0, 1, 2, 3, and 4, it is discrete. An example of a continuous random variable X_c is the distance of a shooter's bullet hole from the bull's eye. As this distance can take any value, X_c is continuous.

Two important functions of a random variable are the *cumulative distribution function* (CDF) and the *probability density function* (PDF).

The cumulative distribution function, $F(x)$, of a random variable X is given by

$$F(x) = P\left[X(\lambda) \le x\right] \tag{3.24}$$

where $P[X(\lambda) \le x]$ is the probability that the value $X(\lambda)$ taken by the random variable X is less than or equal to the quantity x.

The cumulative distribution function $F(x)$ has the following properties:

1. $0 \le F(x) \le 1$
2. $F(x_1) \le F(x_2)$ if $x_1 \le x_2$
3. $F(-\infty) = 0$
4. $F(+\infty) = 1$

The probability density function, $f(x)$, of a random variable X is the derivative of $F(x)$ and thus given by

$$f(x) = \frac{dF(x)}{dx} \qquad (3.25)$$

The probability density function $f(x)$ has the following properties:

1. $f(x) \geq 0$ for all values of x

2. $\int_{-\infty}^{\infty} f(x)dx = 1$

Further, from Eqs. (3.24) and (3.25), we have

$$F(x) = \int_{-\infty}^{x} f(z)dz \qquad (3.26)$$

The function within the integral is not shown as a function of x because, as per Eq. (3.24), x is defined here as a fixed quantity. It has been arbitrarily shown as a function of z, where z has the same dimension as x, $f(z)$ being the same PDF as $f(x)$. Some texts, however, show it equivalently as a function of x, with the understanding that x is used in the generalized sense.

The following example will help in clarifying the concepts behind the PDF, $f(x)$, and the CDF, $F(x)$. In Fig. 3.4a, a four-level *pulse amplitude modulation* (PAM) signal is shown. The amplitude of each pulse is random and equally likely to occupy any of the four levels. Thus, if a random variable X is defined as the signal level v, and $P(v = x)$ is the probability that $v = x$, then

$$P(v = -3) = P(v = -1) = P(v = +1) = P(v = +3) = 0.25 \qquad (3.27)$$

With this probability information, we can determine the associated CDF, $F_{4L}(v)$. For example, for $v = -1$

$$F_{4L}(-1) = P(v \leq -1) = P(v = -3) + P(v = -1) = 0.5 \qquad (3.28)$$

In a similar fashion, $F_{4L}(v)$ for other values of v may be determined. A plot of $F_{4L}(v)$ versus v is shown in Fig. 3.4b.

The PDF $f_{4L}(v)$ corresponding to $F_{4L}(v)$ can be found by differentiating $F_{4L}(v)$ with respect to v. The derivative of a step of amplitude V is a pulse of value V. Thus, since the steps of $F_{4L}(v)$ are of value 0.25

$$f_{4L}(-3) = f_{4L}(-1) = f_{4L}(+1) = f_{4L}(+3) = 0.25 \qquad (3.29)$$

A plot of $f_{4L}(v)$ versus v is shown in Fig. 3.4c.

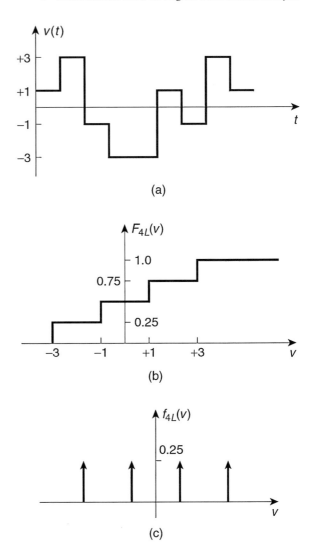

Fig. 3.4 A four-level PAM signal and its associated CDF and PDF. (**a**) 4-level PAM signal. (**b**) Cumulative distribution function (CDF) associated with levels of 4-level PAM signal. (**c**) Probability distribution function (PDF) associated with levels of 4-level PAM signal

3.3.2   *The Average Value, the Mean Squared Value, and the Variance of a Random Variable*

The *average value* or *mean, m,* of a random variable X, also called the *expectation* of X, is denoted either by $\overline{X}$ or $E(x)$. For a discrete random variable, X_d, where n is the total number of possible outcomes of values $x_1, x_2, \ldots, x_n$, and where the probabilities of the outcomes are $P(x_1), P(x_2), \ldots, P(x_n)$, it can be shown that

$$m = \overline{X_d} = E\left(X_d\right) = \sum_{i=1}^{n} x_i P\left(x_i\right) \tag{3.30}$$

For a continuous random variable X_c, with PDF $f_c(x)$, it can be shown that

$$m = \overline{X_c} = E\left(X_c\right) = \int_{-\infty}^{\infty} x \cdot f\left(x\right) dx \tag{3.31}$$

and that the *mean square value*, $\overline{X_c^2}$, or $E\left(X_c^2\right)$ is given by

$$\overline{X_c^2} = E\left(X_c^2\right) = \int_{-\infty}^{\infty} x^2 f\left(x\right) dx \tag{3.32}$$

In Fig. 3.5 is shown an arbitrary PDF of a continuous random variable. A useful number to help in evaluating a continuous random variable is one which gives a measure of how widely spread its values are around its mean m. Such a number is the root mean square (rms) value of $(X-m)$s and is called the *standard deviation σ* of X.

The square of the standard deviation, σ^2, is called the variance of X and is given by

$$\sigma^2 = E\left[\left(X-m\right)^2\right] = \int_{-\infty}^{\infty} \left(x-m\right)^2 f\left(x\right) dx \tag{3.33}$$

The relationship between the variance σ^2 and the mean square value $E(X^2)$ is given by

$$\begin{aligned}
\sigma^2 &= E\left[\left(X-m\right)^2\right] \\
&= E\left[X^2 - 2mX + m^2\right] \\
&= E\left(X^2\right) - 2mE\left(X\right) + m^2 \\
&= E\left(X^2\right) - m^2
\end{aligned} \tag{3.34}$$

We note that for the average value $m = 0$, the variance $\sigma^2 = E(X^2)$.

Fig. 3.5 A probability distribution function (PDF) of a continuous random variable

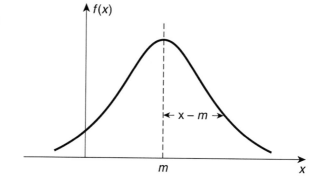

3.3.3 The Gaussian Probability Density Function

The *Gaussian* or, as it's sometimes called, the *normal* PDF [1, 2] is very important to the study of wireless transmission, being the function most often used to describe *thermal noise*. Thermal noise is the result of thermal motions of electrons in the atmosphere, resistors, transistors, etc. and is thus unavoidable in communication systems. The Gaussian probability density function, $f(x)$, is given by

$$f(x) = \frac{1}{\sqrt{2\pi\sigma^2}} e^{-(x-m)^2/2\sigma^2} \tag{3.35}$$

where m is as defined in Eq. (3.30) and σ as defined in Eq. (3.33). When $m = 0$ and $\sigma = 1$, the *normalized Gaussian probability density function* is obtained. A graph of the Gaussian PDF is shown in Fig. 3.6a.

The CDF corresponding to the Gaussian PDF is given by

$$F(x) = P\big[X(\lambda) \le x\big] = \int_{-\infty}^{x} \frac{e^{-(z-m)^2/2\sigma^2}}{\sqrt{2\pi\sigma^2}} dz \tag{3.36}$$

When $m = 0$, the *normalized Gaussian cumulative distribution function* is obtained, being given by

$$F(x) = \int_{-\infty}^{x} \frac{e^{-z^2/2\sigma^2}}{\sqrt{2\pi\sigma^2}} dz \tag{3.37}$$

A graph of the Gaussian cumulative distribution function is shown in Fig. 3.6b. In practice, since the integral in Eq. (3.37) is not easily determined, it is normally evaluated by relating it to the well-known and numerically computed function, the *error function*. The error function of v is defined by

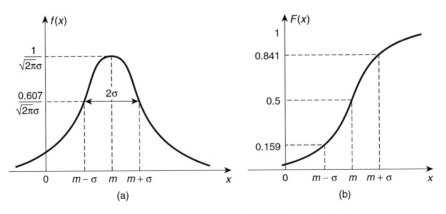

Fig. 3.6 The Gaussian random variable. (**a**) Density function. (**b**) Distribution function

$$erf(v) = \frac{2}{\sqrt{\pi}} \int_0^v e^{-u^2} du \qquad (3.38)$$

and it can be shown that erf (0) = 0 and erf(∞) = 1.

The function $[1 - erf(v)]$ is referred to as the *complementary error function*, erfc (v). Noting that $\int_0^v = \int_0^\infty - \int_v^\infty$, we have

$$erfc(v) = 1 - erf(v)$$

$$= 1 - \left[\frac{2}{\sqrt{\pi}} \int_0^\infty e^{-u^2} du - \frac{2}{\sqrt{\pi}} \int_v^\infty e^{-u^2} du \right]$$

$$= 1 - \left[erfc(\infty) - \frac{2}{\sqrt{\pi}} \int_v^\infty e^{-u^2} du \right] \qquad (3.39)$$

$$= \frac{2}{\sqrt{\pi}} \int_v^\infty e^{-u^2} du$$

Tabulated values of erfc(v) are only available for positive values of v.

Using the substitution $u \equiv x / \sqrt{2}\sigma$, it can be shown [1] that the Gaussian CDF $F(x)$ of Eq. (3.37) may be expressed in terms of the complementary error function of Eq. (3.39) as follows

$$F(x) = 1 - \frac{1}{2} erfc\left(\frac{x}{\sqrt{2}\sigma}\right) \quad \text{for } x \geq 0 \qquad (3.40a)$$

$$= \frac{1}{2} erfc\left(\frac{|x|}{\sqrt{2}\sigma}\right) \text{ for } x \leq 0 \qquad (3.40b)$$

3.3.4 The Rayleigh Probability Density Function

The propagation of wireless signals through the atmosphere is often subject to multipath fading. Such fading will be described in detail in Chap. 4. Multipath fading is best characterized by the *Rayleigh* PDF [1]. Other phenomena in wireless transmission are also characterized by the Rayleigh PDF, making it an important tool in wireless analysis. The Rayleigh probability density function $f(r)$ is defined by

$$f(r) = \frac{r}{\alpha^2} e^{-r^2/2\alpha^2}, \quad 0 \leq r \leq \infty \qquad (3.41a)$$

$$= 0, \quad r < 0 \qquad (3.41b)$$

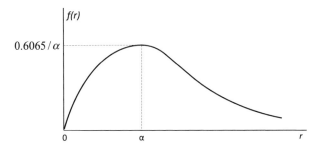

Fig. 3.7 The Rayleigh probability density function

and hence the corresponding CDF, that is, the probability that $R(\lambda)$ does not exceed a specified level r, is given by

$$F(r) = P\left[R(\lambda) \le r\right] = 1 - e^{-r^2/2\alpha^2}, \quad 0 \le r \le \infty \qquad (3.42a)$$

$$= 0, \quad r < 0 \qquad (3.42b)$$

A graph of $f(r)$ as a function of r is shown in Fig. 3.7. It has a maximum value of $1/(\alpha\sqrt{e})$ which occurs at $r = \alpha$. It has a mean value $\overline{R} = \sqrt{\pi/2} \cdot \alpha$, a mean square value $\overline{R^2} = 2\alpha^2$, and hence, by Eq. 3.34, a variance σ^2 given by

$$\sigma^2 = \left(2 - \frac{\pi}{2}\right)\alpha^2 \qquad (3.43)$$

A graph of $F(r)$ versus $10\log_{10}(r^2/2\alpha^2)$, which is from Feher [3], is shown in Fig. 3.8. If the amplitude envelope variation of a radio signal is represented by the Rayleigh random variable R, then the envelope has a mean square value of $\overline{R^2} = 2\alpha^2$, and hence the signal has an average power of $\overline{R^2}/2 = \alpha^2$. Thus, $10\log_{10}(r^2/2\alpha^2)$, which equals $10\log_{10}(r^2/2) - 10\log_{10}(\alpha^2)$, represents the decibel difference between the signal power level when its amplitude is r and its average power. From Fig. 3.8 it will be noted that for signal power less than the average power by 10 dB or more, the distribution function $F(r)$ decreases by a factor of 10 for every 10 dB decrease in signal power. As a result, whenever a fading radio signal exhibits this behavior, the fading is described as Rayleigh fading.

3.3.5   Thermal Noise

White noise [1] is defined as a random signal whose power spectral density is constant, i.e., independent of frequency. True white noise is not physically realizable since constant power spectral density over an infinite frequency range implies

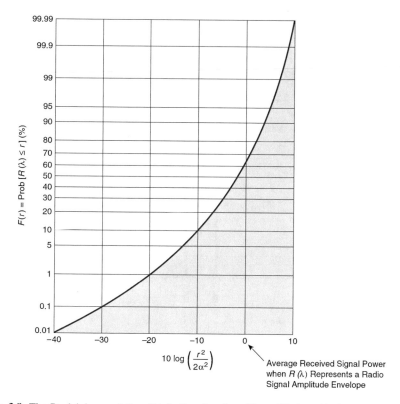

Fig. 3.8 The Rayleigh cumulative distribution function. (From [3], by with the permission of the author)

infinite power. However, thermal noise, which as indicated earlier has a Gaussian PDF, has a power spectral density that is relatively uniform up to frequencies of about 1000 GHz at room temperature (290°K) and up to of about 100 GHz at 29°K [4]. Thus, for the purpose of practical communications analysis, it is regarded as white. A simple model for thermal noise is one where the two-sided power spectral density $G_n(f)$ is given by

$$G_n(f) = \frac{N_0}{2} \mathrm{W / Hz} \tag{3.44}$$

where N_0 is a constant.

In a typical wireless communications receiver, the incoming signal and accompanying thermal noise is normally passed through a symmetrical bandpass filter centered at the carrier frequency f_c to minimize interference and noise. The width of the bandpass filter, W, is normally small compared to the carrier frequency. When this is the case, the filtered noise can be characterized via its so-called narrowband representation [1]. In this representation, the filtered noise voltage, $n_{nb}(t)$, is given by

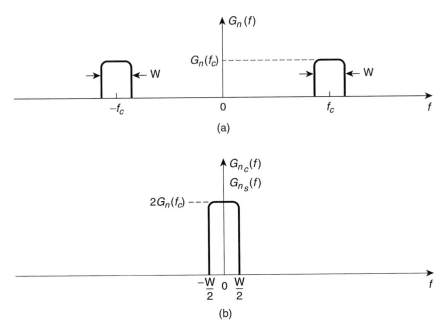

Fig. 3.9 Spectral density relationships associated with narrowband representation of noise. (**a**) Power spectral density of bandpass filtered thermal noise. (**b**) Power spectral density of narrowband noise representation components n_c and n_s

$$n_{nb}(t) = n_c(t)\cos 2\pi f_c t - n_s(t)\sin 2\pi f_c t \tag{3.45}$$

where $n_c(t)$ and $n_s(t)$ are Gaussian random processes of zero mean value, of equal variance and, further, independent of each other. Their power spectral densities, $G_{n_c}(f)$ and $G_{n_s}(f)$, extend only over the range $-W/2$ to $W/2$ and are related to $G_n(f)$ as follows

$$G_{n_c}(f) = G_{n_s}(f) = 2G_n(f_c + f) \tag{3.46}$$

The relationship between these power spectral densities is shown in Fig. 3.9. This narrowband noise representation is very useful in the study of carrier modulation methods.

3.3.6 Noise Filtering and Noise Bandwidth

In a receiver, a received signal contaminated with thermal noise is normally filtered to minimize the noise power relative to the signal power prior to demodulation. If, as shown in Fig. 3.10, the input two-sided noise spectral density is $N_0/2$, the transfer

Fig. 3.10 Filtering of white noise

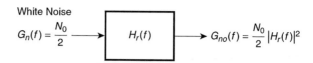

function of the real filter is $H_r(f)$, and the output noise spectral density is $G_{no}(f)$, then, by Eq. (3.23), we have

$$G_{no}\left(f\right)=\frac{N_0}{2}\left|H_r\left(f\right)\right|^2 \tag{3.47}$$

and thus, the normalized noise power at the filter output, P_o, is given by

$$
\begin{aligned}
P_o &= \int_{-\infty}^{\infty} G_{no}\left(f\right)df \\
&= \frac{N_0}{2}\int_{-\infty}^{\infty}\left|H_r\left(f\right)\right|^2 df
\end{aligned}
\tag{3.48}
$$

A useful quantity to compare the amount of noise passed by one receiver filter versus another is the filter *noise bandwidth* [1]. The noise bandwidth of a filter is defined as the width of an ideal brick-wall (rectangular) filter that passes the same average power from a white noise source as does the real filter. In the case of a real low-pass filter, it is assumed that the absolute values of the transfer functions of both the real and brick-wall filters are normalized to one at zero frequency. In the case of a real bandpass filter, it is assumed that the brick-wall has the same center frequency as the real filter, f_c say, and that the absolute values of the transfer functions of both real and brick-wall filters are normalized to one at f_c.

For an ideal brick-wall low-pass filter of two-sided bandwidth B_n and $|H_{bw}(f)| = 1$

$$P_o = \frac{N_0}{2}B_n \tag{3.49}$$

Thus, from Eqs. (3.48) and (3.49), we determine that,

$$B_n = \int_{-\infty}^{\infty}\left|H_r\left(f\right)\right|^2 df \tag{3.50}$$

Figure 3.11 shows the transfer function $H_{bw}(f)$ of a brick-wall filter of noise bandwidth B_n superimposed on the transfer function $H_r(f)$ of a real filter.

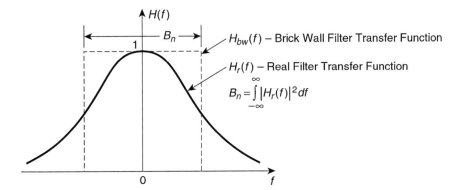

Fig. 3.11 Low-pass filter two-sided noise bandwidth, B_n

3.4 Summary

In this chapter spectral analysis of non-periodic functions and relevant statistical methods were reviewed. As was shown, spectral analysis permits the characterization of signals in the frequency domain and provides the relationship between frequency domain and time domain characterizations. Noise and propagation anomalies are random processes leading to uncertainty in the integrity of a recovered signal. By employing statistical methods, however, computation of the fidelity of the recovered baseband signal is possible in terms of the probability that it's in error. The study of non-light-of-sight propagation analysis set out in Chap. 4, the basic principles of wireless modulation covered in Chap. 5, and the multiple access techniques outlined in Chap. 7 will apply several of the tools presented here.

References

1. Taub H, Schilling D (1971) Principles of communication systems. Mc-Graw-Hill, Inc, New York
2. Cooper GR, McGillem CD (1971) Probabilistic methods of signal and system analysis. Holt. Rinehart and Winston, Inc, New York
3. Feher K (ed) (1987) Advanced digital communications. Prentice-Hall, Inc., Upper Saddle River
4. Members of the Technical Staff, Bell Telephone Laboratories (1971) Transmission systems for communications. Revised 4th edn. Bell Telephone Laboratories, Inc, Winston-Salem, North Carolina

Chapter 4
The Mobile Wireless Path

4.1 Introduction

A mobile wireless system communicates between a base station and mobile units via the propagation of radio waves over a *path*. In an ideal world, the path would be free of obstruction, i.e., have a line of sight (LOS) between transmitter and receiver and attenuate the transmitted signal by a fixed amount across the transmitted signal spectrum, resulting in a predictable and undistorted signal level at the receiver input. Such attenuation is referred to as *free space loss*. In the real mobile world, however, paths are rarely LOS. Rather, they are mainly non-line of sight (NLOS), and the intervening topography and atmospheric conditions normally result in a received signal that deviates significantly from ideal. Changes to the received signal over and above free space loss are referred to as fading.

A typical base station to mobile unit link is shown in Fig. 4.1. As will be seen in succeeding sections, the maximum length of a wireless path for reliable communications varies depending on (a) the propagation frequency; (b) the antenna heights; (c) terrain conditions between the sites, in particular, the clearance or lack thereof between the direct signal path and ground obstructions; (d) atmospheric conditions over the path; and (e) the radio equipment and antenna system electrical parameters. For mobile systems operating in the 1 to 10 GHz bands, paths for reliable communications, depending on the general surrounding topography, can be up to several kilometers in length. For those operating in the 24 to 100 GHz bands, however, path lengths are much more restricted, being typically less than a kilometer in length, much of this restriction being due to higher free space loss as a result of the higher frequency and higher obstruction losses. In this chapter we will examine propagation in an ideal environment and then study the various types of fading and how such fading impacts path reliability. As antennas provide the means to efficiently launch and receive radio waves, a brief review of their characteristics is in order as the first step in addressing ideal propagation.

© Springer Nature Switzerland AG 2020
D. H. Morais, *Key 5G Physical Layer Technologies*,
https://doi.org/10.1007/978-3-030-51441-9_4

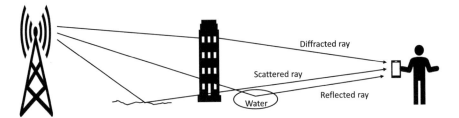

Fig. 4.1 Typical base station to mobile unit link

4.2   Antennas

4.2.1   Introduction

For 5G base stations, antennas provide either omnidirectional coverage, coverage over a sector, typically 120° in azimuth, or highly directional coverage to individual mobile units. For PMP mobile links, the mobile unit antennas are normally omnidirectional in sub-6 GHz bands but can be directional in the millimeter wave bands. For fixed wireless access links, the remote antenna, if outdoor, is normally directional. A base station omnidirectional antenna is usually a vertical stack of dipole antennas, referred to as a *stacked dipole* or *collinear array*. A base station sector antenna, on the other hand, is usually a linear array of dipole antennas combined electrically so as to produce directivity and is referred to as a flat plane or planar array antenna. Many antenna characteristics are important in designing PMP wireless broadband systems. The most important of these will be reviewed followed by a brief description of those antennas commonly used in such systems.

4.2.2   Antenna Characteristics

Antenna gain is the most important antenna characteristic. It is a measure of the antenna's ability to concentrate its energy in a specific direction relative to radiating it isotropically, i.e., equally in all directions. The more concentrated the beam, the higher the gain of the antenna. It can be shown [1] that a transmitting antenna that concentrates its radiated energy within a small beam, i.e., a directional antenna, has a gain G in the direction of maximum intensity with respect to an isotropic radiator of

$$G(\mathrm{dB}) = 10 \cdot \log_{10}\left(\frac{4\pi A_e}{\lambda^2}\right) \tag{4.1}$$

where A_e = effective area of the antenna aperture and λ = wavelength of the radiated signal.

The antenna's physical area A_p is related to its effective area by the following relationship

$$A_e = \eta A_p \tag{4.2}$$

where η is the efficiency factor of the antenna.

Equation (4.1) indicates that a directional antenna's gain is function of the square of the frequency ($\lambda f = c$); thus doubling the frequency increases the gain by 6 dB. It is also a function of the area of the aperture. Thus, for a parabolic antenna, for example, the gain is a function of the square of the antenna's diameter, and so doubling the diameter increases the gain by 6 dB. The efficiency factor η in Eq. (4.2) accounts for the fact that antennas are not 100 percent efficient. In the case of a parabolic antenna, this is because not all the total incident power on the radiator is radiated forward as per theory. Some of it is lost to "spillover" at edges, some is misdirected because the radiator surface is not manufactured perfectly to the desired shape, and some is blocked by the presence of the feed radiator. The nominal value of η for a parabolic antenna is 0.55 (55% efficient). The operation of an antenna in the receive mode is the inverse of its operation in the transmit mode. It therefore comes as no surprise that its receive gain, defined as the energy received by the antenna compared to that received by an isotropic absorber, is identical to its transmit gain. Figure 4.2 shows a plot of antenna gain versus angular deviation from its direct axis, which is referred to as its *pole* or *boresight*. It shows the main lobe where most of the power is concentrated. It also shows side lobes and back lobes that can cause interference into or from other wireless systems in the vicinity.

The *beamwidth* of an antenna is closely associated with its gain. The higher the gain of the antenna, the narrower the width of the beam. Beamwidth is measured in radians or degrees and is usually defined as the angle that subtends the points at which the peak field power is reduced by 3 dB. Figure 4.2 illustrates this definition. The narrower the beamwidth, the more interference from external sources including nearby antennas is minimized. This improvement comes at a price, however, as narrow beamwidth antennas require precise alignment. This is because a very small shift in alignment results in a measurable decrease in the transmitted power directed at a receiving antenna and the received power from a distant transmitting antenna.

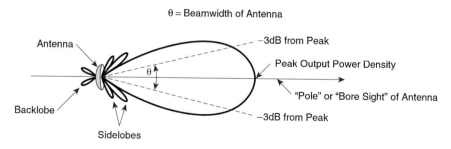

Fig. 4.2 Antenna gain versus angular deviation from its pole

The *front-to-back ratio* is another important antenna characteristic. It is defined as the ratio, usually expressed in dBs, of its maximum gain in the forward direction to the maximum gain in the region of its backward direction, the latter being the maximum back lobe gain. Unfortunately, the front-to-back ratio of an antenna in a real installation can vary widely from that in a purely free space environment. This variation is due to foreground reflections in a backward direction of energy from the main transmission lobe by objects in or near the lobe. These backward reflections can reduce the free space only front-to-back ratio by 20 to 30 dB.

The polarization of an antenna refers to the alignment of the electric field in the radiated wave. Thus, in a *horizontally polarized* antenna, the electric field is horizontal, and in a *vertically polarized* antenna, the electric field is vertical. When a signal is transmitted in one polarization, a fraction of the signal may be converted to the other polarization due to imperfections in the antenna and the path. The ratio of the power received in the desired polarization to the power received in the undesired polarization is referred to as the *cross-polarization discrimination*. Cross polarization typical varies from about 25 to 40 dB depending on the antennas and the path.

Finally, we note that an antenna-related parameter often regulated for wireless systems is the *equivalent isotropically radiated power* (*EIRP*). This power is the product of the power supplied to the transmitting antenna, P_t, say, and the gain of the transmitting antenna, G_{ta} say. Thus, we have

$$EIRP = P_t \cdot G_{ta} \qquad (4.3)$$

4.2.3   Typical Point-to-Multipoint Broadband Wireless Antennas

The stacked dipole or collinear array can be used at a base station to provide omnidirectional coverage. It should be noted that it does not actually provide true omnidirectional coverage, since such coverage would be equal in all directions and hence isotropic. What it provides is omnidirectional coverage in the horizontal (or close to horizontal) plane only. In the vertical plane, its radiated energy diminishes as it propagates in a more and more vertical direction until it's essentially zero in the truly vertical direction. The simplest omnidirectional antenna is a single half-wave dipole. Its basic form is shown in Fig. 4.3a and its horizontal and vertical radiation pattern shown in Fig. 4.3b. The theoretical gain of this antenna is only 2.15 dB. Fortunately, by simply stacking identical dipoles into a single structure in the vertical plane and feeding them simultaneously with a phase-sensitive network, higher gain is achieved. When stacked vertically, the resulting antenna is the stacked dipole. The gain achieved over a single dipole is a function of the spacing between the dipoles and, depending on the number of dipoles stacked, peaks for a dipole center-to-center spacing of between approximately 0.9 and 1.0 wavelengths. A stacked dipole with four dipole elements has a maximum gain of approximately 8 dB, and one with eight elements has a gain of about 11 dB.

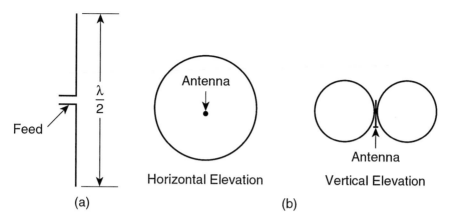

Fig. 4.3 Half-wave dipole antenna. (**a**) Basic structure. (**b**) Radiation pattern

Fig. 4.4 Typical dipole
planar array

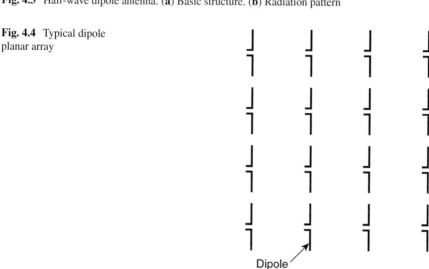

The *planar array* or flat panel antenna is typically used at a base station to provide sectorized coverage. One of its key attributes is its flat profile. One physical realization of a planar array antenna is dipole antennas arranged in a planar array. The elements may be either single polarized or dual polarized where there are two dipoles per element perpendicular to each other. A planar array provides two dimensions of control, allowing a beam, highly directive in both the horizontal and vertical coordinates, to be produced. Though each dipole in the array has an omnidirectional radiation pattern, they are cleverly connected together via a network that results in directivity in a forward direction along a line perpendicular to the plane of the array. Figure 4.4 shows a 4 × 4 single polarized array.

Another physical realization of the planar antenna is "patches" arranged in a planar array. Each patch is a rectangular sheet of metal, approximately half the operating frequency wavelength on each side, mounted over, but electrically insulated from, a sheet of metal serving as a ground plane. The feed point is along one edge of the rectangle, and for cross-polarized operation, the two feed points are located on perpendicular sides. Figure 4.5 is a graphical representation of a 2×2 patch antenna array.

Mobile unit antennas, as indicated in Sect. 4.2.1 above, are normally omnidirectional in the sub-6 GHz range. They are typically, depending on the operating frequency, half-wave dipoles, quarter wave monopoles (a monopole is a single conductor antenna that's reliant on a nearby ground plane), or complex variations thereof. In the millimeter range, however, directional antennas are preferable, and such antennas are usually planar arrays. Since there is a need for directivity regardless of the orientation of the unit, multiple arrays are used in different locations around the handset case.

4.3   Free Space Propagation

As indicated above, the propagation of a signal over a wireless path is affected by both atmospheric anomalies and the intervening terrain. Absent of any such interfering effects, we have signal loss only as a result of free space. Free space loss is defined as the loss between two isotropic antennas in free space. Consider the point source shown in Fig. 4.6 radiating isotropically a signal of power P_t into free space. As the surface area of a sphere of radius d is $4\pi d^2$, then the radiated power density $p(d)$ on a sphere of radius d, centered on the point source, is given by

$$p(d) = \frac{P_t}{4\pi d^2} \tag{4.4}$$

If a receiving antenna with an effective area A_{ef} is located on the surface of the sphere, then the power received by this antenna, P_r, will be equal to the power density on the sphere times the effective area of the antenna, i.e.,

Fig. 4.5 4×4
Patch antenna

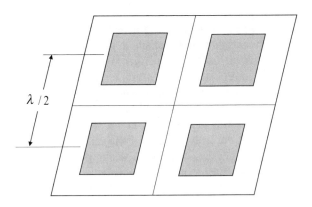

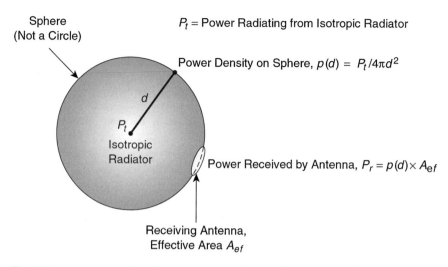

Fig. 4.6 Power density radiated from an isotropic radiator and power received by an antenna

$$P_r = \frac{P_t A_{ef}}{4\pi d^2} \tag{4.5}$$

To determine free space loss, we need to know the power received by an isotropic antenna. But, in order to know this, Eq. (4.5) indicates that we need to know the effective area of an isotropic antenna. By definition, the gain G of an isotropic antenna is 1. Substituting this value of gain into Eq. (4.1) gives

$$A_{ef} = \frac{\lambda^2}{4\pi} \tag{4.6}$$

Substituting Eq. (4.6) into Eq. (4.5), we determine that P_r, the power receiver by an isotropic antenna, is related to P_t, the power transmitted by an isotropic radiator, by the following relationship

$$P_r = \frac{P_t}{\left(\dfrac{4\pi d}{\lambda}\right)^2} \tag{4.7}$$

The denominator of the right-hand side of Eq. (4.7) represents the free space loss, L_{fs}, experienced between the isotropic antennas. It is usually expressed in its logarithmic form, i.e.,

$$L_{fs}(\text{dB}) = 20\log_{10}\left(\frac{4\pi d}{\lambda}\right) \tag{4.8}$$

Substituting in Eq. (4.8), the well-known relationship $\lambda f = c$, where c is the speed of electromagnetic propagation and equals 3×10^8 m/s for free space transmission, we get

$$L_{fs}\,(\mathrm{dB}) = 32.4 + 20\log_{10} f + 20\log_{10} d \qquad (4.9a)$$

where f is the transmission frequency in MHz; d is the transmission distance in km and

$$L_{fs}\,(\mathrm{dB}) = 36.6 + 20\log_{10} f + 20\log_{10} d \qquad (4.9b)$$

where f is the transmission frequency in MHz; d is the transmission distance in miles.

To get a sense of the magnitude of free space loss typical of wireless broadband links, consider a 2 GHz path, 3 km long, and a 25 GHz path 0.3 km long. For the 2 GHz path, the free space loss would be 108 dB, and for the 25 GHz path, the loss would be 110 dB.

Thus, even at one tenth the distance of the 2 GHz path, the 25 GHz path suffers slightly more loss than the 2 GHz path.

4.4   Line-of-Sight Non-faded Received Signal Level

With a relationship to determine free space loss, one is now able to also set out a relationship for the determination of the receiver input power P_r in a typical wireless link assuming no loss to fading or obstruction. Starting with the transmitter output power P_t, one simply accounts for all the gains and losses between the transmitter output and the receiver input. These gains and losses, in dBs, in the order incurred are:

L_{tf} = loss in transmitter antenna feeder line (coaxial cable or waveguide depending on frequency)
G_{ta} = gain of the transmitter antenna
L_{fs} = free space loss
G_{ra} = gain of the receiver antenna
L_{rf} = loss in the receiver antenna feeder line

Thus

$$P_r = P_t - L_{tf} + G_{ta} - L_{fs} + G_{ra} - L_{rf} \qquad (4.10a)$$

$$= P_t - L_{sl} \qquad (4.10b)$$

where $L_{sl} = L_{tf} - G_{ta} + L_{fs} - G_{ra} + L_{rf}$ is referred to as the *section loss* or *net path loss*.

Example 4.1 Computation of received input power

A PMP link has the following typical parameters:

Path length, $d = 3$ km
Operating frequency, $f = 2$ GHz
Transmitter output power, $P_t = 40.0$ dBm (10 W)
Loss in transmitter antenna feeder line, $L_{tf} = 2$ dB
Transmitter antenna gain, $G_{ta} = 10$ dB
Receiver antenna gain, $G_{ra} = 2$ dB
Loss in receiver antenna feeder line, $L_{rf} = 0.2$ dB

What is the receiver input power?

Solution

By Eq. (4.9a) the free space loss, L_{fs}, is given by

$$L_{fs} = 32.6 + 20\log_{10} 2{,}000 + 20\log_{10} 3 = 108.2 \text{dB}$$

Thus, by Eq. (4.10a), the receiver input power, P_r, is given by

$$P_r = 40 - 2 + 10 - 108.2 + 2 - 0.2 = -58.4 \text{dBm}$$

For a fixed LOS wireless link such as may be found in a fixed wireless access system, received input power P_r resulting from free space loss only is normally designed to be higher than the minimum receivable or threshold level R_{th} for acceptable probability of error. This power difference is engineered so that if the signal fades, due to various atmospheric and terrain effects, it will fall below its threshold level for only a small fraction of time. This built-in level difference is called the link *fade margin* and thus given by

$$\text{Fade margin} = P_r - R_{th} \tag{4.11}$$

4.5 Fading Phenomena

The propagation of radio signals through the atmosphere is affected by terrain features in or close to its path in addition to the atmospheric effects. Potential propagation affecting terrain features are trees, hills, sharp points of projection, reflection surfaces such as ponds and lakes, and man-made structures such as buildings and towers. These features can result in either reflected or diffracted signals arriving at the receiving antenna. Potential propagation affecting atmospheric effects, on relatively short paths, includes rain, water vapor, and oxygen.

4.5.1   *Fresnel Zones*

The proximity of terrain features to the direct path of a radio signal impacts their effect on the composite received signal. *Fresnel zones* are a way of defining such proximity in a very meaningful way. In the study of the diffraction of radio signals, the concept of Fresnel zones is very helpful. Thus, before reviewing diffraction, an overview of Fresnel zones is in order. The first Fresnel zone is defined as that region containing all points from which a wave could be reflected such that the total length of the two segment reflected path exceeds that of the direct path by half a wavelength, $\lambda/2$, or less. The nth Fresnel zone is defined as that region containing all points from which a wave could be reflected such that the length of the two segment reflected path exceeds that of the direct path by more than $(n - 1)\lambda/2$ but less than or equal to $n\lambda/2$. The boundary of a Fresnel zone surrounding the direct path in the plane of the path is an ellipsoid whereas it is a circle in the plane perpendicular to the path. Figure 4.7 shows the first and second Fresnel zone boundaries on a line-of-sight path. The perpendicular distance F_n from the direct path to the outer boundary of the nth Fresnel zone is approximated by the following equation:

$$F_n = \left[\frac{n\lambda d_{n1} d_{n2}}{d} \right]^{\frac{1}{2}} \tag{4.12}$$

where d_{n1} = distance from one end of path to point where F_n is being determined; d_{n2} = distance from other end of path to point where F_n is being determined; d = length of path = $d_{n1} + d_{n2}$; λ, d_{n1}, d_{n2} and d are measured in identical units
   Specifically, F_1 is given by

$$F_1 = 17.3 \left[\frac{d_{11} d_{12}}{fd} \right]^{\frac{1}{2}} \text{ m} \tag{4.13a}$$

where d_{11}, d_{12}, and d are in km and f is in GHz,
   and given by

$$F_1 = 72.1 \left[\frac{d_{11} d_{12}}{fd} \right]^{\frac{1}{2}} \text{ feet} \tag{4.13b}$$

where d_{11}, d_{12} and d are in miles and f is in GHz.
   Thus, for a 1 km path operating at 2 GHz, the maximum value of F_1, which occurs at the middle of the path, is, by Eq. (4.13a), 6.1 m. However, for a typical 1 km, 25 GHz path, F_1 maximum is only 1.73 m.
   Most of the power that reaches the receiver is contained within the boundary of the first Fresnel zone (more on this in the next section). Thus, terrain features that lie outside this boundary, with the exception of highly reflective surfaces, do not, in

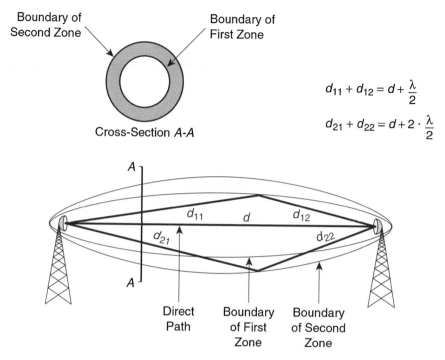

$$d_{11} + d_{12} = d + \frac{\lambda}{2}$$

$$d_{21} + d_{22} = d + 2 \cdot \frac{\lambda}{2}$$

Fig. 4.7 First and second Fresnel zone boundaries

general, significantly affect the level of the received signal. For a signal reflected at this boundary that acquires a $\lambda/2$ phase shift as a result of the reflection, its total phase shift at the receiving antenna relative to the direct signal is λ, and hence the reflected signal is additive to the direct signal. However, for a signal that experiences zero phase shift, as is possible with a vertically polarized reflected signal with a large incident angle, the total relative phase shift at the receiving antenna is $\lambda/2$ and hence results in a partial cancellation of the direct signal. The maximum increase possible in signal strength resulting from a reflected signal is 6 dB. However, the maximum loss possible is in theory an infinite number of decibels and in practice can exceed 40 dB.

4.5.2 Reflection

The reflection of a radio signal occurs when that signal impinges on a surface which has very large dimensions compared to the wavelength of the signal. Reflections typically occur from the surface of the earth, particularly liquid surfaces, and from buildings. When reflection occurs, the reflected signal arriving at a receiving antenna is combined with the direct signal to form a composite signal that, depending on the strength and phase of the reflected signal, can significantly degrade the desired

Fig. 4.8 Reflected signal

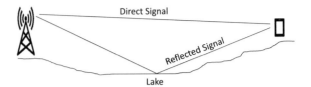

signal. Figure 4.8 shows a radio path with a direct signal and a reflected signal off a lake. The strength of a reflected signal at a receiving antenna depends on the directivity of both the transmitting and receiving antenna, the heights of these antennas above the ground, the length of the path, and the strength of the reflected signal relative to the incident signal at the point of reflection. The phase of a reflected signal at a receiving antenna is a function of the reflected signal path length and the phase shift, if any, incurred at the point of reflection.

The strength and phase of a reflected signal relative to the incident signal at the point of reflection are strongly influenced by the composition of the reflecting surface, the curvature of the surface, the amount of surface roughness, the incident angle, and the radio signal frequency and polarization. If reflections occur on rough surfaces, they usually do not create a problem as the incident and reflected angles are quite random. Reflections from a relatively smooth surface, however, depending on the location of the surface, can result in the reflected signal being intercepted by the receiving antenna. The transfer function of a reflective surface is referred to as its *reflection coefficient, R*. For a highly reflective surface (a specular reflection plane), vertically polarized signals show more frequency dependence than horizontally polarized ones. A method to calculate effective surface reflection coefficient is given in Sect. 6.1.2.4.1 of ITU Recommendation ITU-R P.530-17 [2].

4.5.3   Diffraction

So far in explaining propagation effects, it has been tacitly assumed that the energy received by a radio antenna travels as a beam from transmitting to receiving antenna that's just wide enough to illuminate the receiving antenna. This is not exactly the case, however, as waves propagate following the *Huygens' principle*. Huygens showed that propagation occurs along a wavefront, with each point on the wavefront acting as a source of a secondary wavefront known as a wavelet, with a new wavefront being created from the combination of the contributions of all the wavelets on the preceding front. Importantly, secondary wavelets radiate in all directions. However, they radiate strongest in the direction of the wavefront propagation and less and less as the angle of radiation relative to the direction of the wavefront propagation increases until the level of radiation is zero in the reverse direction of wavefront propagation. The net result is that, as the wavefront moves forward, it spreads out, albeit with less and less energy on a given point as that point is removed further and further from the direct line of propagation. This sounds like bad news.

The signal intercepted at the receiving antenna, however, is the sum of all signals directed at it from all the wavelets created on all wavefronts as the wave moves from transmitter to receiver. At the receiver, signal energy from some wavelets tend to cancel signal energy from others depending on the phase differences of the received signals, these differences being generated as a result of the different path lengths. The net result is that, in a free space, unobstructed environment, half of the energy reaching the receiving antenna is canceled out.

Signal components that have a phase difference of $\lambda/4$ or less relative to the direct line signal are additive. Signals with a phase difference between $\lambda/4$ and $\lambda/2$ are subtractive. In an unobstructed environment, all such signals fall within the first Fresnel zone. In fact, the first Fresnel contains most of the power that reaches the receiver. Consider now what happens when an obstacle exists in a radio path within the first Fresnel zone. Clearly, under this condition, the amount of energy intercepted at the receiving antenna will differ from that intercepted if no obstacle were present. The cause of this difference, which is the disruption of the wavefront at the obstruction, is called *diffraction*. If an obstruction is raised in front of the wave so that a direct path is just maintained, the power reaching the receiver will be reduced, whereas the simplistic narrow beam model would suggest that full received signal would be maintained. A positive aspect of diffraction is that if the obstruction is further raised, so that it blocks the direct path, signal will still be intercepted at the receiving antenna, albeit at a lesser and lesser level as the height of the obstruction increases further and further. Under the simplistic narrow beam model, one would have expected complete signal loss.

Figure 4.9a shows "unobstructed" free space propagation, where path clearance is assumed to exceed several Fresnel zones. Figure 4.9b shows diffraction around an obstacle assumed to be within the region of the first Fresnel zone but not blocking the direct path. Figure 4.9c shows diffraction around an obstacle blocking the direct path. For simplicity an expanding wave that has progressed partially down the path is shown in all three depictions, to the point of obstruction in the case of the obstructed paths. All wavelets on the wavefront shown in the unobstructed case are able to radiate a signal that falls on the receiving antenna. In the case of the obstructed paths, however, the size of the wavefronts and hence the number of wavelets radiating signals that fall on the receiving antenna are decreased.

Like loss due to reflection, diffraction loss is a function of the nature of the diffracting terrain, varying from a minimum for a single knife-edge-type obstacle to a maximum for a smooth earth-type obstacle. Much research and analysis of diffraction loss have been conducted over several decades leading to well-accepted formulae [3] for its estimation. For the situation where the clearance between an obstacle and the direct path is equal to F_1, the first Fresnel zone clearance at the obstacle location, then there is, in fact, a received signal gain. This gain varies from about 1.5 dB for the knife edge case to 6 dB for the smooth earth case. When the clearance decreases to $0.6\ F_1$, the received signal strength is unaffected by the obstruction, regardless of its type. For the grazing situation, where the clearance between the obstacle and the direct path is reduced to zero, diffraction loss varies from approximately 6 dB for the knife-edge case to approximately 15 dB for the smooth earth case.

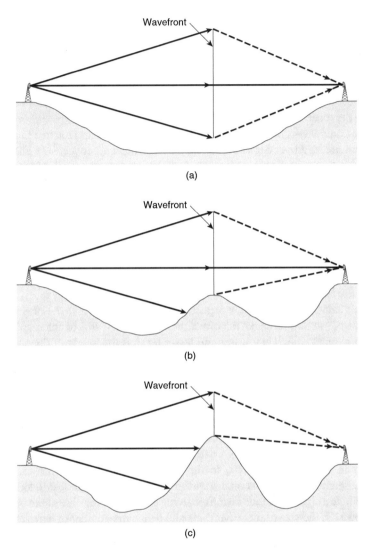

Fig. 4.9 Graphical representation of diffraction. (**a**) Unobstructed path. (**b**) Path with obstacle that doesn't block direct path. (**c**) Path with obstacle that blocks direct path

4.5.4 Scattering

Scattering occurs when a radio signal hits a large rough surface or objects with dimensions that are small compared to signal wavelength. The energy reflected from such surfaces tends to be spread out (scattered) in many directions, and thus some of this energy may impinge on the receiver. Figure 4.10 below shows a

Fig. 4.10 Scattering scenario

scattering scenario. Typical urban scatterers include rough ground, foliage, street signs, and lamp posts. Often, in a mobile radio environment, the received signal is stronger than that which would have been expected as a result of reflection and diffraction due to the added energy received via scattering. Of course, the phase of the received scattered signal could be such as to reduce overall signal strength.

4.5.5   Rain Attenuation and Atmospheric Absorption

As a radio signal propagates down its path, it may find itself subjected to, in addition to the possible effects of reflection, diffraction, and scattering, the attenuating effects of rain and absorption by atmospheric gasses, primarily water vapor and oxygen.

Raindrops attenuate a radio signal by absorbing and scattering the radio energy, with this effect becoming more and more significant as the wavelength of the signal decreases towards the size of the raindrops. Figure 4.11, which is from Rogers et al. [4], shows attenuation in dB/km versus frequency for several rain conditions. As can be seen from the figure, attenuation increases with propagation frequency and rain intensity. Attenuation due to a heavy cloudburst at 6 GHz is about 1 dB/km, whereas at 24 GHz, it is about 20 dB/km. The width of cloudburst cells tends to be on the order of kilometers, so maximum attenuation due to rain at 24 GHz for a 1 km. path would be approximately 20 dB. In general, rain attenuation is not a major problem below about 10 GHz. However, as can be seen from Fig. 4.11, this situation changes measurably as the frequency increases and normally must be factored in at frequencies of 10 GHz and above. The results in Fig. 4.11 are theoretically derived for spherical raindrops. Real raindrops, however, are somewhat flattened as they fall through the atmosphere. As a result, they have a smaller size in the vertical plane than in the horizontal one. This in turn leads to the attenuation of a vertically polarized wave being less than that of a horizontally polarized one.

Figure 4.12, which is from ITU Recommendation ITU-R P.676-5 [5], shows attenuation in dB/km due to water vapor and dry air as a function of frequency in a low to moderate humidity region. From the figure it will be observed that attenuation due to water vapor has a first peak at approximately 22 GHz with a value of about 0.18 dB/km. It then declines as the frequency increases up to about 31 GHz, after which it increases back to about 0.18 dB/km in the region of 60 GHz. Attenuation due to dry air is below 0.01 dB/km for frequencies up to 20 GHz and

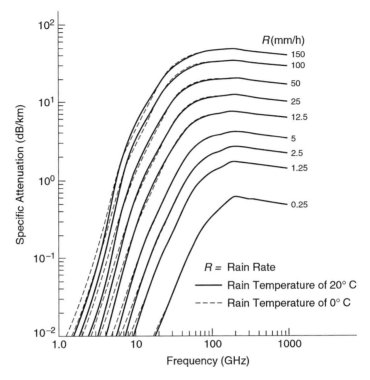

Fig. 4.11 Attenuation due to rain. (From [4], with the permission of the Communications Research Center, Canada)

then increases somewhat exponentially to a peak of approximately 15 dB/km in the region of 60 GHz. Paths close to 60 GHz, unfortunately, have the worst of two worlds. Rain attenuation can approach 40 dB/km and is additive to the dry air attenuation of approximately 15 dB/km. As a result, PMP wireless systems operating at frequencies in this region and in locations subject to any appreciable rainfall are usually limited in path length to much less than 1 km.

4.5.6 Penetration Loss

When a mobile unit or a fixed unit including its antenna is indoors, the received signal suffers additional loss over and above that suffered as a result of outdoor effects as a result of the signal having to penetrate either a wall, or a window, or a door, or a combination thereof. This added loss is thus referred to as *penetration loss* and can be quite large.

"Different materials commonly used in building construction have very diverse penetration loss characteristics. Common glass tends to be relatively transparent

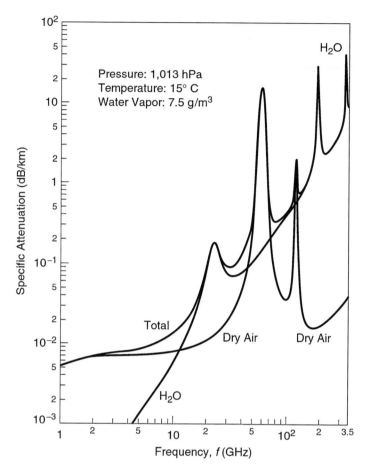

Fig. 4.12 Specific attenuation due to atmospheric gasses. (From [5], with the permission of the ITU)

with a rather weak increase of loss with higher frequency due to conductivity losses. Energy efficient glass commonly used in modern buildings or when renovating older buildings is typically metal-coated for better thermal insulation. This coating introduces additional losses that can be as high as 40 dB even at lower frequencies. Materials such as concrete or brick have losses that increase rapidly with frequency" [6].

Typical building exteriors are constructed of many different materials, e.g., concrete, sheet rock, wood, brick, glass, etc. Thus, depending on the location of the receiver within the building, its final signal may be a combination of several signals traversing several different materials with different penetration loss properties. Empirical data on building entry loss can be found in the ITU report P.2346-3 [7].

4.6   Signal Strength Versus Frequency Effects

4.6.1   Introduction

Fading in PMP systems is the variation in strength of a received radio signal due normally to terrain and atmospheric effects in the radio's path as discussed in Sections 4.5 above. Fading is normally broken down into two main types based on the impact of fading on the signal spectrum. If fading attenuates a signal uniformly across its frequency band, fading is referred to as *flat fading*. Fading due to rain or atmospheric gasses is typically flat fading. If fading results in varying attenuation across the signal's frequency band, such fading is called *frequency selective fading*. These two types of fading can occur separately or together. It cannot be predicted with any accuracy exactly when fading is likely to occur.

4.6.2   Frequency Selective Fading

If a radio path passes over highly reflective ground or water or a building surface and the antenna heights allow, in addition to a direct path, a reflected path between the antennas, reflection-induced fading is likely and, under certain circumstances, can be substantial. Reflection-induced fading is frequency selective. The reflected signal will have, at any instant, a path length difference and hence a time delay, τ say, relative to the direct path that's independent of frequency. However, this delay, measured as a phase angle, varies with frequency. Thus, if the propagated signal occupies a band between frequencies f_1 and f_2 say, then the relative phase delay in radians at frequency f_1 will be $2\pi f_1 \tau$, the relative phase delay at f_2 will be $2\pi f_2 \tau$, and the difference in relative phase delay between the component of the reflected signal at f_1 and that at f_2 will be $2\pi(f_2 - f_1)\tau$. Because the reflected signal will have a phase shift relative to the direct signal that's a function of frequency, then, when both these signals are combined, a different resultant signal relative to the direct signal will be created for each frequency. Depending on the value of τ and the signal bandwidth $f_2 - f_1$, significant differences can exist in the composite received signal amplitude and phase as a function of frequency as compared to the undistorted direct signal. Further, the closer the reflected signal is in amplitude to the direct signal, the larger the maximum signal cancellation possible, this occurring when the signals are 180° out of phase.

   To aid in understanding how a reflected signal can result in a received signal that is frequency selective, consider the system depicted in Fig. 4.13. Given a time delay between the direct signal, S_d, and reflected signal, S_r, of τ s, then the phase delay between S_d and S_r is equal to $2\pi f\tau$. Let us assume that the received signals are in phase at the frequency f_1. Then the phase delay between the two signals at f_1 must be equal to $n.2\pi$, where n is an integer. Thus

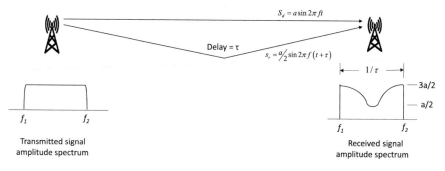

Fig. 4.13 Frequency selective transmission

$$n.2\pi = 2\pi f_1 \tau \qquad (4.14)$$

and hence

$$f_1 = n / \tau. \qquad (4.15)$$

Let us further assume that the received signals are next in phase at f_2. Then the phase delay between the two signals at f_2 must be equal to $(n + 1)2\pi$, where n is an integer. Thus

$$(n+1)2\pi = 2\pi f_2 \tau \qquad (4.16)$$

and hence

$$f_2 = (n+1) / \tau. \qquad (4.17)$$

Then, by Eqs. (4.15) and (4.17),

$$f_2 - f_1 = 1 / \tau \qquad (4.18)$$

If we assume, as shown in Fig. 4.12, that the amplitude of the received direct signal is a and that of the reflected signal is $a/2$, then the combined received signal amplitude at both f_1 and f_2 is equal to $3a/2$. At $(f_1 + f_2)/2$, however, the frequency centered between f_1 and f_2, the signals are out of phase, and thus the combined received signal amplitude is $a/2$.

4.7   NLOS Path Analysis Parameters

For point-to-multipoint NLOS paths, path analysis consists of estimating, for a given distance d from the base station, the following parameters:

- *Mean path loss* of all possible paths of length *d*.
- *Shadowing*, i.e., variation about this mean due to the impact of varying surrounding terrain features from one path to the next.
- *Multipath fading*, i.e., distortion and/or amplitude variation of the composite received signal resulting from the varying delay in time between the various received signals. These varying time delays result from movement in the surrounding environment, for example, moving vehicles, people walking, and swaying foliage.
- *Doppler shift fading*, i.e., distortion and/or amplitude variation of the composite received signal resulting from variations in the frequencies of received signal components as a result of motion.

4.7.1   Mean Path Loss

As indicated earlier, for LOS paths with good clearance, signal loss between a transmitter and receiver antenna is free space loss, given by

$$L_{fs} = 32.4 + 20\log f + 10\log d^2 \tag{4.19}$$

where *f* is in MHz and *d* is in km, and loss in dBs.

The equation above indicates that the loss is proportional to the square of distance. For NLOS (shadowed) paths, however, and not surprisingly, this relationship does not hold, since the composite received signal is sum of a number of signals attenuated by diffraction, reflection, scattering, and atmospheric effects, if any. Rather, it has been found empirically that in such situations, the ensemble average loss $\overline{L_p}(d)$ (in dBs) of all possible paths on circle of radius *d* from the base station can be stated as follows:

$$\overline{L_p}(d) = L_p(d_0) + 10\log\left(\frac{d}{d_0}\right)^n \tag{4.20}$$

where d_0 is a close-in reference distance around base station free of obstruction; $L_p(d_0)$ is loss at d_0 and is either measured or computed as free space loss; *n* is the *path loss exponent*.

For mobile access cells, d_0 is often specified as 100 m for macro-cells and as low as 1 m for microcells. Path loss exponent *n* can vary from 2 in free space up to about 5 in heavily shadowed urban environments. Note that in a shadowed environment, doubling the path distance increases path loss by $3n$ dB. Thus, for $n = 2$ (free space), loss increases by 6 dB, but for $n = 5$ say, loss increases by 15 dB! Models have been devised that estimate *n* in terms of terrain conditions for a specific frequency and specific base station and remote station antenna heights. To use these models for other frequencies and other antenna, heights correction terms are added to the loss determined by applying *n* in the basic path loss equation. Since $\overline{L_p}(d)$ helps characterize signal strength over "large distances" (a change in d of a few wavelengths won't change $\overline{L_p}(d)$ in a measurable way), $\overline{L_p}(d)$ is referred to as a *large-scale loss* component.

4.7.2 Shadowing

As indicated in the preceding, $\overline{L_p}(d)$ is the ensemble average loss for all paths on the circle of distance d from the base station. The actual loss at each location, however, can be significantly different from this average as the surrounding terrain features can vary greatly from location to location. Measurements have shown that for any particular value d, the path loss $L_p(d)$ is a random variable having a Gaussian (normal) distribution distributed about $\overline{L_p}(d)$. Thus

$$L_p(d) = \overline{L_p}(d) + X_\sigma \tag{4.21}$$

where X_σ denotes a zero-mean Gaussian random variable expressed in dBs, i.e., a log-normal variable, with standard deviation σ (also expressed in dBs). The standard deviation σ is the rms value of X and gives a measure of how widely spread the values of X are around its mean. Figure 4.14 below shows the general format of the probability density function $F(x)$ of $L_p(d)$ which is the probability density function of X_σ superimposed on $\overline{L_p}(d)$.

Figure 4.15 [8] shows a scatter plot of measured path loss versus distance for a mobile cell in the Seattle area. Note the spread of measurements around the mean for any given distance.

The close-in reference distance d_0, the path loss exponent n, and the standard deviation σ statistically describe a path loss model for an arbitrary location having a specific base station-remote station separation of value d. In practice, σ is computed from measured data or assigned a value based on the coverage area topography. In general, in cluttered cities and hilly areas of heavy tree density, σ tends to be high, whereas in rural areas with relatively flat terrain and low tree density, it tends to be low. Since X_σ helps characterize signal strength over "large distances" (a change in location of a few wavelengths won't change X_σ in a significant way), X_σ, like $\overline{L_p}(d)$, is a large-scale loss component.

For a loss of $L_p(d)$, transmitter power of P_t, and total transmitter and receiver antenna gain G_a, the received signal level $P_r(d)$ is given by

$$P_r(d) = P_t + G_a - L_p(d) = P_t + G_a - \overline{L_p}(d) - X_\sigma \tag{4.22}$$

In path design, the average received signal level, i.e., $P_t + G_a - \overline{L_p}(d)$, is normally designed to be above a given receiver threshold level, x_0, by a so-called shadow margin, M_s say. For this margin, there is an associated probability that the received signal level $P_r(d)$ will be greater than or equal to the threshold level x_0, i.e., $P[P_r(d) \geq x_0]$. This probability is the probability that $X_\sigma \leq M_s$ and hence given by

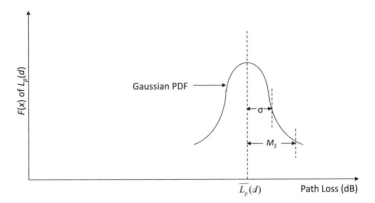

Fig. 4.14 General format of the probability density function of $L_p(d)$

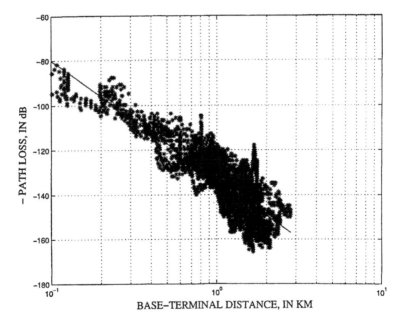

Fig. 4.15 Scatter plot of path loss versus distance for a mobile cell in the Seattle area. (From [8], with the permission of the IEEE)

$F(M_s)$, where $F(M_s)$ is the cumulative distribution of X_σ. Since X_σ is a Gaussian random variable of zero mean, $F(M_s)$ is given by Eq. 3.40a. Thus

$$P\left[P_r(d) \geq x_0\right] = P\left[X_\sigma \leq M_s\right] = F(M_s) = 1 - \frac{1}{2}erfc\left(\frac{M_s}{\sqrt{2}\sigma}\right) \qquad (4.23a)$$

$$= 1 - Q\left(\frac{M_s}{\sigma}\right) \tag{4.23b}$$

For example, when $\sigma = 8$ dB and $M_s = 10$ dB, then $P[P_r(d) \geq x_0] = 0.94$, i.e., 94%.

Note: When $M_s = \sigma$, then $P[P_r(d) \geq x_0]$ is 84%.

The shadow margin M_s lets us determine what percentage of paths at a given distance d from the base station will have a large-scale loss equal to or less than $\overline{L_p}(d) + M_s$ and hence a received signal level $P_r(d)$ equal to or greater than a given threshold value x_0 say. Interest, however, is normally more in F_u, the fraction of circular coverage area of radius d from the BS having a received signal level equal to or greater than a given threshold x_0. It can be shown [9] that F_u is a function of σ, the path exponent n, and the probability of coverage at d, $P[P_r(d) \geq x_0]$.

Figure 4.16 [9] gives a family of curves for determining F_u. In the figure, $P[P_r(d) \geq x_0]$ is shown as $P_{x0}(R)$. From this figure it can be determined that, for example, if $n = 4$, $\sigma = 8$, and if the probability of coverage at the boundary is 0.75, then the fraction of total area with signal above the threshold, F_u, is 0.9.

4.7.3 Path Loss Models

A number of path loss models applicable to 5G mobile systems have been introduced by major organizations or groups thereof in the recent past. These models are all more complex than models introduced for 3G and to some extent 4G systems.

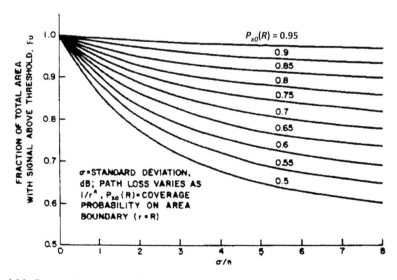

Fig. 4.16 Curves relating F_u, the fraction of circular area of radius d with received signal $P_r(d)$ above threshold x_0, as a function of the probability of $P_r(d)$ being above x_0 at the cell boundary. (From [9], with the permission of the IEEE)

It is not possible to describe them in a few equations as was previously the case, and thus relevant equations will not be set out here. They all typically provide both LOS and NLOS versions, and input parameters typically include:

- A general description of the surrounding topography, e.g., indoor, dense urban, urban, and rural
- Base station antenna height
- Mobile unit height
- Ground-level distance between base station and mobile unit
- Direct distance between the base station antenna and mobile unit antenna
- An assumed shadow fading standard deviation in dBs
- Transmission frequency

Among these new models are:

1. The ITU model: ITU-R M.2412-0 [10]
2. The 3GPP model: 3GPP TR 38.901 [11]
3. Model by an ad hoc group of companies and universities (5GCM) [12]
4. The European Union sponsored METIS model [13]
5. The European Union sponsored Millimeter-Wave Based Mobile Radio Access Network for Fifth Generation Integrated Communications (mmMAGIC) model [14]

4.7.4   Multipath Fading

At a receiver in a NLOS path, the received signal typically consists of multiple versions of transmitted signal as a result of reflection, diffraction, and scattering. These signals all arrive at receiver at slightly different times and from different directions, given that they traverse different paths. As a result, they have different amplitudes and phases and add up vectorially at receiver to form a composite received signal. A channel with such multiple signals is called a multipath channel. Figure 4.17 illustrates such a channel.

If amplitudes/phases of signals received over a multipath channel remain fixed, then a composite received signal of fixed amplitude/phase results. Depending on (a) the time delay between various signals and (b) the signal modulating pulse duration, differing signals can cause the composite signal to be distorted, diminished in amplitude, or a mix of both. If the amplitudes and phases of the received signals vary, however, then the composite signal's amplitude and phase also vary. This results in varying signal distortion, amplitude fluctuations, or a mix of both, depending again on (a) the time delay between various signals and (b) the signal modulating pulse duration (more on this later). These signal variations are referred to as multipath fading. In a NLOS path with a stationary remote unit, the received signal may still suffer multipath fading due to movement in the surrounding environment, for example, moving vehicles, people walking, and swaying foliage. Even the

Fig. 4.17 Multipath channel

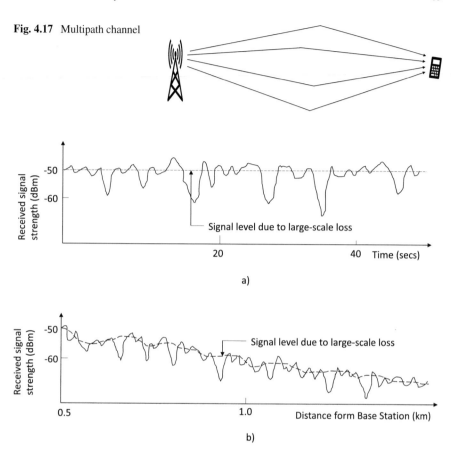

Fig. 4.18 Small-scale fading superimposed on large-scale loss. (**a**) Received signal strength at a fixed UE showing small-scale fading superimposed on received signal level due to large-scale loss. (**b**) Received signal strength at mobile UE showing small-scale fading superimposed on received signal level due to large-scale loss

smallest movements cause variations in amplitudes and phases of reflected signals, and the rapidity of these movements translates into rapidity of induced fading. Such rapid fading is therefore referred to as *small-scale fading*. Figure 4.18a below depicts small-scale fading as a function of time at a fixed remote unit, and Fig. 4.18b depicts small-scale fading as a function of displacement from the base station.

Multipath fading described above is called *Rayleigh fading* if in the received signal there is no line-of-sight (nonfading) component. In such a signal, the composite envelope is described statistically by the Rayleigh probability density function (Sect. 3.3.4). When there is a dominant, nonfading, LOS component accompanied by random multipath components arriving at different angles, the fading is called *Rician fading*, as the composite envelope is described statistically by the Rician pdf [9]. The Rician distribution is often described in terms of K, where K

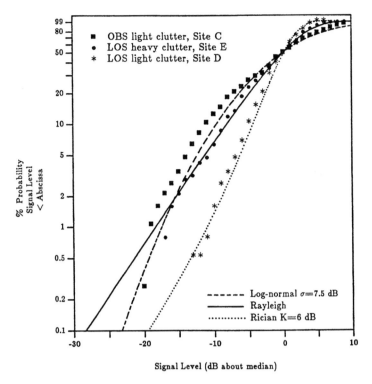

Fig. 4.19 Raleigh, Rician, and log-normal distribution functions. (From [15], with the permission of the IEEE)

is the ratio of power in the dominant component to the mean power in the variable components. As the dominant signal becomes smaller and smaller relative to the variable components, i.e., as K becomes smaller and smaller, the composite signal's characteristics resemble more and more one with a Raleigh distribution, and in the extreme ($K = 0$), the Rician distribution transforms to Raleigh distribution. With Rayleigh fading, received signal power may vary by as much as 30–40 dB as a result of small-scale movement, on the order of a wavelength or less, taking place over a period on the order a second.

Figure 4.19 [15] shows the Raleigh, Rician for $K = 6$ dB, and log-normal cumulative distribution functions. We note that, with Rician fading, a fade of 20 dB or greater only occurs for 0.1% of the time, whereas with Raleigh fading, a fade of 20 dB or greater occurs 0.7% of the time, i.e., seven times as often.

NLOS multipath fading can be flat or frequency selective. As indicated above, fading is considered flat if the channel has a relatively constant gain and linear phase over the bandwidth of the transmitted signal, whereas it is considered frequency selective if the channel has significantly varying gain and phase over the bandwidth of the transmitted signal. Figure 4.20 shows signal spectral density versus frequency. In Fig. 4.20a the transmission channel experiences flat fading as the spectral

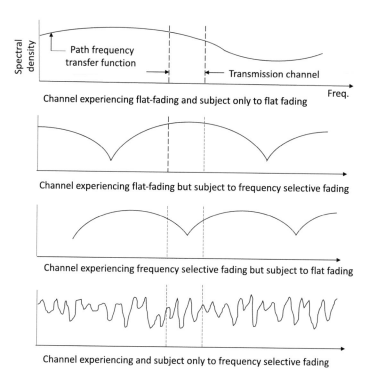

Channel experiencing flat-fading and subject only to flat fading

Channel experiencing flat-fading but subject to frequency selective fading

Channel experiencing frequency selective fading but subject to flat fading

Channel experiencing and subject only to frequency selective fading

Fig. 4.20 Channels experiencing flat fading and frequency selective fading

density across a spectrum much broader than the transmission channel is fairly smooth; thus, even though the location of peaks and valleys are likely to change with time, the variation across the transmission channel is likely to remain fairly flat. In Fig. 4.20b the peaks and valleys of the broader spectrum are quite sharp. Thus, though the transmission channel is shown experiencing flat fading, it is subject to frequency selective fading as the broader spectrum shifts with time. In Fig. 4.20c the situation is the reverse of that in Fig. 4.20b. The transmission channel is shown experiencing frequency selective fading but subject to flat fading. Figure 4.20d shows a transmission channel subject only to frequency selective fading.

A measure of the range of frequencies over which a channel can be considered to begin approaching being "flat" is called the channel's *coherence bandwidth B_c*. Coherence bandwidth is defined in terms of the channels *rms delay spread σ_τ*. A channel's *mean excess delay, rms delay spread*, and *maximum excess delay spread* (*X* dB) are multipath channel parameters that can be derived from its power delay profile. A channel's power delay profile is a plot of relative received signal power as a function of excess delay with respect to the first signal component that arrives at a receiver when a single impulse is transmitted. For a typical wireless channel, the received signal usually consists of several discrete components. Figure 4.21 is a representation of a highly cluttered outdoor power delay profile.

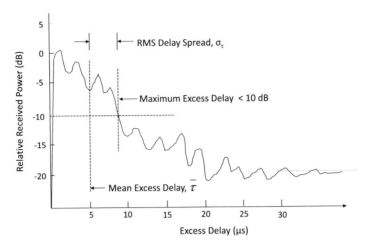

Fig. 4.21   Power delay profile

If mean excess delay of all components when weighted by their power levels is given by $\bar{\tau}$ and if mean excess delay squared of all components when weighted by their power levels is given by $\overline{\tau^2}$, then rms delay spread σ_τ is given by

$$\sigma_\tau = \sqrt{\left(\overline{\tau^2}\right) - \left(\bar{\tau}\right)^2} \tag{4.24}$$

Example 4.2   RMS delay spread computation

For the profile in Fig. 4.22:

Mean excess delay $\bar{\tau} = \left[(1 \times 0) + (.5 \times 5) + (.35 \times 10)\right] / (1 + .5 + .35) = 3.24\,\mu s$

Mean excess delay squared $\overline{\tau^2} = \left[(1 \times 0^2) + (.5 \times 5^2) + (.35 \times 10^2)\right] / 1.85 = 25.68\,\mu s^2$

Thus, rms delay spread σ_τ given by

$$\sigma_\tau = \sqrt{25.68 - (3.24)^2} = \sqrt{15.18} = 3.90\,\mu s$$

Depending on terrain, distance, antenna directivity, and other factors, values of σ_τ can vary from tens of nanoseconds to microseconds.

Let's return now to coherence bandwidth B_c. This bandwidth is not an exact quantity but rather a frequency domain indication of flatness. One popular definition of B_c is:

$$B_c \approx 1 / 5\sigma_\tau \tag{4.25}$$

Thus, for example, with this definition, if $\sigma_\tau = 1\,\mu s$, then $B_c = 200$ kHz.

Fig. 4.22 RMS delay profile

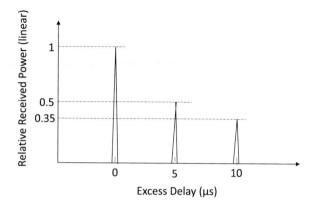

If a signal is modulated with symbols of length T_s, and so has a Nyquist bandwidth of $B_s = 1/T_s$, then the signal is said to undergo flat fading if:

$$B_s \ll B_c \tag{4.26a}$$

i.e.,

$$T_s \gg 5\sigma_\tau \tag{4.26b}$$

Flat fading causes little or no ISI but results in increased bit error rate (BER) due to a decreased signal-to-noise ratio. Additional margin is typically added to protect against multipath flat fading. Diversity systems are also highly effective in reducing the negative effects of multipath flat fading.

We turn now to frequency selective fading. A signal is said to undergo frequency selective fading if:

$$B_s \geq B_c \tag{4.27a}$$

i.e.,

$$\sigma_\tau \geq T_s / 5 \tag{4.27b}$$

What this indicates is that a measurable amount of a delayed pulse combines with a non-delayed succeeding pulse to create ISI. Techniques such as OFDM combat frequency selective fading in NLOS PMP systems. OFDM is discussed in Chap. 7.

3GPP developed three *extended ITU channel models* [16] for LTE evaluation:

- Extended Pedestrian A (EPA): Low delay spread model employed in (a) an urban environment with a fairly small cell size, i.e., < than about 1 km, and (b) suburban environment with cell size up to about 2 km.
- Extended Vehicular A (EVA): Typical urban vehicular model.
- Extended Typical Urban (ETU): Extreme urban model.

Table 4.1 Extended ITU channel model rms delay spread and maximum excess delay

Relative delay spread	Channel model	rms delay spread σ (ns)	Maximum excess delay (ns)
Low	Extended Pedestrian A (EPA)	43	410 @ −20.8 dB rel. power
Medium	Extended Vehicular A (EVA)	357	2510 @ −16.9 dB rel. power
High	Extended Typical Urban (ETU)	991	5000 @ −7.0 dB rel. power

The rms delay spread and maximum excess delay for these extended ITU models are given in Table 4.1 above to give a sense of the range of values for these delays that can be expected:

4.7.5   Doppler Shift Fading

In addition to amplitude/phase signal changes caused by changes in path geometries, a received composite signal is also affected by motion in the surrounding environment. Each signal received by a moving object or each multipath wave that's reflected off a moving object experiences an apparent shift in frequency, this shift being called a *Doppler shift*. This shift is proportional to the speed with which the object moves and a function of the direction of motion. The result is that the spectrum of the received signal is broadened. The effect of this broadening is again fading, specifically small-scale fading, in the form of signal distortion, amplitude fluctuations, or a mix of both, depending on the relationship between (a) the amount of broadening and (b) the signal modulation pulse duration. Like multipath fading, Doppler shift fading is rapid hence it being classified as small-scale fading. Doppler spread B_d is a measure of spectral broadening of a channel caused by the rate of change of the channel over time, resulting primarily from the mobile's motion in case of a mobile unit and motion of reflectors and scatterers in the case of a fixed remote unit. Figure 4.23 demonstrates the impact of speed and direction on apparent wavelength.

A received signal of carrier frequency f_c, suffering Doppler spread, will have components in the range $f_c - f_d$ to $f_c + f_d$, where f_d is called the *Doppler shift*. It can be shown that for a vehicle moving at speed v receiving a signal of wavelength λ, the maximum value of f_d is given by

$$\left(f_d\right)_{max} = v / \lambda$$

Thus

$$B_d = 2\left(f_d\right)_{max} = 2v / \lambda = vf_c / 540 \text{ Hz} \tag{4.28}$$

where v is in km/h, and f_c is in MHz

Coherence time T_c is the time domain equivalent of Doppler spread. Coherence time, like coherence bandwidth, is not an exact quantity, but a popular definition is:

$$T_c = \frac{0.423}{(f_d)_{max}} = 0.423\frac{\lambda}{v} = 0.423\frac{c}{vf_c} = \frac{457}{vf_c} \text{ s.} \tag{4.29}$$

where c is speed of light, v is velocity in km/h, and f_c is in MHz.

If a signal is modulated with symbols of length T_s and so has a Nyquist bandwidth of $B_s = 1/T_s$, then a signal is said to undergo Doppler shift-induced *slow fading* if:

$$B_s \gg B_d \tag{4.30a}$$

i.e.

$$T_s \ll T_c \tag{4.30b}$$

In a slow fading environment, the effects of Doppler spread on the quality of received signal are negligible.

A signal is said to undergo Doppler shift-induced *fast fading* if:

$$T_s > T_c \tag{4.31a}$$

$$B_s < B_d \tag{4.31b}$$

In a fast fading channel, the impulse response of the channel changes rapidly within the duration of the transmitted symbol. This causes *frequency dispersion*, referred also to as *time selective fading*, which leads to signal distortion.

Typical maximum Doppler spreads for mobile broadband wireless applications as per Eq. 4.27 are given in Table 4.2 below:

In a NLOS path simultaneously experiencing Rayleigh fading and Doppler shift fading, then how rapidly the channel fades is a function of the relative speed between the transmitter and receiver. The faster the speed, the more rapid the fading.

Fig. 4.23 Impact of speed on apparent wavelength

Table 4.2 Maximum Doppler spreads

Carrier freq. (GHz)	Speed (km/h)	Max. Doppler spread (kHz)
2	60	0.22
2	120	0.44
4	60	0.44
4	120	0.89
40	60	4.44
40	120	8.89

4.7.6   Doppler Shift and Multipath Fading Analysis

In the following example, Doppler shift and multipath fading on an OFDM signal are analyzed for a given set of conditions.

Example 4.3 Doppler shift and multipath fading analysis

Conditions:

- Mobile subscribers traveling at speeds of up to 120 km/h
- Operating Frequency: 2.0 GHz
- Assumed channel model: Extended Typical Urban
- OFDM subcarrier bandwidth = 15 kHz

Analysis:

- Doppler shift fading:

  - Relevant signal bandwidth, B_s, is the bandwidth of the individual subcarriers that constitute the OFDM signal, i.e., B_s = 15 kHz.
  - Transmission frequency = 2.0 GHz and maximum speed = 120 km/h. Therefore, from earlier table, Doppler spread, B_d = 0.44 kHz.
  - Thus, $B_s \gg B_d$, *Doppler shift fading slow*, and signal impairment from Doppler shift minimal.

- Multipath fading:

  - Channel mode: ETU. Thus, from earlier table rms delay spread is 0.991 μs.
  - Therefore, coherence bandwidth B_c = $1/5\sigma$ = $1/(5 \times 0.991 \times 10^{-6})$ Hz = 202 kHz.
  - Thus $B_s \ll B_c$.
  - Hence, *multipath fading flat*.

4.8   Millimeter Wave Communications

Broadly stated, one of the goals of 5G NR is to operate in frequencies from below 1 GHz to as high as 100 GHz. In Rel. 15 and early Rel. 16, however, NR supports operation in two specified frequency ranges:

FR1: In 3GPP TS 38.104 Rel. 15, ver. 15.1.0, bands starting as low as 450 MHz and ending as high as 6 GHz, referred to as the sub-6 GHz range. In ver. 15.5.0, however, upper end of range extended to 7.125 GHz. Wider range unchanged in early Rel. 16 versions.

FR2: In Rel. 15 and early Rel. 16 versions, bands starting as low as 24.25 GHz and ending as high as 52.60 GHz, commonly referred to as the millimeter wave (mmwave) range.

Since both the operating frequencies and the channel bandwidths of NR millimeter wave signals are much higher than those in FR1, it is not surprising that propagation models for millimeter wave bands require more than simply scaling those developed for the FR1. Such millimeter wave encompassing models have already been referenced in Sect. 4.7.3 above. Good overviews of 5G millimeter wave communications are given in References [17, 18].

Referring to Fig. 4.12, we note that attenuation due to dry air increases rapidly above about 20 GHz, reaching to a peak value of about 15 dB/km at a frequency just above 50 GHz. Attenuation due to water vapor has a first peak value of about 0.2 dB/km at just over 20 GHz, dips a bit to just under 0.1 dB/km at 30 GHz, and then recommences raising to about 0.4 dB/km at 100 GHz. Referring to Fig. 4.11, we see that rain results in significant attenuation above 10 GHz, the value being about 25 dB/km at 25 GHz and continues increasing to about 50 dB/km at 100 GHz.

As with atmospheric effects, diffraction loss is measurably higher at millimeter wave frequencies compared to sub-6GHz frequencies. The reason for this is readily apparent if we consider the respective first Fresnel zone boundaries. Such boundaries are inversely proportional to the square root of the frequency. The first Fresnel zone boundary of a 25 GHz signal is thus 3.2 times smaller than that of a 2.5 GHz signal assuming the same path length. Since most of the power that reaches the receiver is contained within the boundary of the first Fresnel zone, then power from a 25 GHz signal will be more quickly blocked by an obstruction than power from a 2.5 GHz one. Given the greater diffraction loss at millimeter wave frequencies compared to sub-6 GHz ones, coupled with higher free space loss, in many situations LOS propagation at millimeter wave frequencies may offer acceptable performance where NLOS propagation may not.

Like diffraction loss, penetration loss also tends to increase with frequency. At 28 GHz, penetration loss was measured by comparing path loss outside by the window and indoors 1.5 m from the window [19]. Median loss was 9 dB for plain glass windows and 15 dB for low-emissivity (low-e) windows. Measurements at 38 GHz [20] found a penetration loss of nearly 25 dB for a tinted glass window and 37 dB for a glass door.

Millimeter wave propagation analysis in a mobile environment is clearly a challenging subject, and additional work needs to be done to achieve results as reliable as those achieved for sub-6 GHz propagation. However, many organizations are addressing this subject, and at some time in the future, competing models will no doubt converge to an acceptable form.

4.9 Summary

A mobile wireless system communicates between a base station and mobile units via the propagation of radio waves over a path. In this chapter we examined propagation in an ideal environment over line-of-sight and non-line-of-sight paths and then studied the various types of fading and how such fading impacts path reliability. As shall be demonstrated in later chapters, this propagation behavior significantly influences the design features of the 5G NR physical layer and the coverage attainable.

References

1. Kraus JD (1950) Antennas. McGraw-Hill Book Company, Inc., New York
2. ITU Recommendation ITU-R P.530-17 (2017) Propagation data and prediction methods required for the design of terrestrial line-of-sight systems. ITU, Geneva
3. ITU Recommendation ITU-R P.526-7 (2001) Propagation by diffraction. ITU, Geneva
4. Rogers DV, Olsen, RL (1976) Calculation of radiowave attenuation due to rain at frequencies of up to 1000 GHz. CRC Report No. 1299, Communications Research Center, Department of Communications, Ottawa
5. ITU Recommendation ITU-R P.676-5 (2001) Attenuation by atmospheric gases. ITU, Geneva Vol. 65, Issue 12, Dec. 2017, pp. 6213–6230
6. Haneda K et al (2016) 5G 3GPP-like channel models for outdoor urban microcellular and macrocellular environments. In: 2016 IEEE international conference on communications workshops (ICCW), May 2016
7. ITU Recommendation ITR-R P.2346-3 (2019) Compilation of measurement data relating to building entry loss. ITU, Geneva
8. Erceg V et al (1999) An empirically based path loss model for wireless channels in suburban environments. IEEE Journal on Selected Areas of Communication 17(7):1205–1211
9. Reudink DO (1974) Properties of Mobile Radio Propagation above 400 MHz. IEEE Transactions on Vehicular Technology vt-23(4):143–159
10. ITU Recommendation ITU-R PM.2412-0 (2017) Guidelines for evaluation of radio interface technologies for IMT-2020. ITU, Geneva
11. 3GPP (2018) 5G; Study on channel model for frequencies from 0.5 to 100 GHz. 3rd Generation Partnership Project (3GPP), TR 38.901 V 14.3.0, Jan 2018
12. 5GCM (2016) 5G channel model for bands up to 100 GHz. Technical Report, Oct 2016
13. METIS (2020) METIS Channel Model. Technical Report METIS2020, Deliverable D1.4 v3, July 2015
14. mmMAGIC (2017) Measurement results and final mmmagic channel models. Technical Report H2020-ICT-671650-mmMAGIC/D2.2, May 2017
15. Rappaport TS (1989) UHF Fading in Factories. IEEE Journal on Selected Areas in Communications 7(1):40–48
16. 3GPP (2017) LTE; Evolved Universal Terrestrial Radio Access (E-UTRA); User Equipment (UE) radio transmission and reception. 3rd Generation Partnership Project (3GPP), TS 36.101 V 14.3.0, Apr 2017
17. Rappaport TS et al (2017) Overwiew of millimeter wave communications for fifth-generation (5G) wireless networks – with a focus on propagation models. IEEE Trans Antennas Propag Spec Issue 5G 65(12): 6213–6230
18. International Wireless Industry Consortium (2019) 5G Millimeter Wave Frequencies and Mobile Networks. IWPC 5G mmWave Mobility White Paper Version 1, June 2019
19. Holma H et al (eds) (2020) 5G technology; 3GPP new radio. Wiley, Hoboken
20. Rodriguez I et al (2015) Analysis of 38 GHz mmwave propagation characteristics of urban scenarios. European wireless 2015, proceedings of 21st European wireless conference, pp 1–8, May 2015

Chapter 5
Digital Modulation: The Basic Principles

5.1 Introduction

The fundamental modulation methods used in 5G are Pi/2 BPSK, QPSK, 16-QAM, 64-QAM, and 256-QAM. These are the methods, therefore, that we are going to study in this chapter, along with some issues involved in their realization. In these methods, baseband signals linearly modulate an RF carrier. Ideally, we would like to cut to the chase and study these methods immediately. This is not what we will do, however. First the fundamentals of baseband transmission techniques will be reviewed, as these form the foundation upon which the techniques involving the linear modulation of an RF carrier are based. This approach will allow us to tractably and with straightforward mathematical analysis determine the spectral and error performance characteristics of the modulation methods under study.

5.2 Baseband Data Transmission

Data transported by communication systems is typically random or very close to random in nature. A stream of random, bipolar, full-length rectangular pulses (a full-length rectangular pulse is one of constant height for the entire pulse duration), where each pulse is equiprobable (probability of being positive is the same as being negative), is in fact a string of nonperiodic pulses. This is so because the value (polarity) of each pulse is entirely independent of all other pulses in the stream. For such a stream, of pulse duration τ, and hence symbol rate $f_s = 1/\tau$ symbols per second, the two-sided amplitude spectral density is that of each of its nonperiodic pulses and thus as given in Eq. (3.3). The absolute value of its normalized positive (real) side is shown in Fig. 5.1. It occupies infinite bandwidth, with the first null at f_s. The spectrum within the first null is normally referred to as the main lobe. In communication systems bandwidth is normally at a premium. Thus, the designer is

© Springer Nature Switzerland AG 2020
D. H. Morais, *Key 5G Physical Layer Technologies*,
https://doi.org/10.1007/978-3-030-51441-9_5

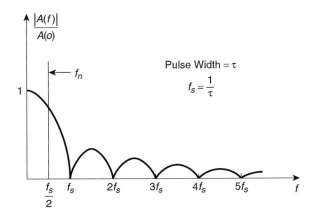

Fig. 5.1 The absolute normalized amplitude spectral density of a train of rectangular pulses

motivated to filter the transmitted signal down to the minimum bandwidth possible without introducing errors into the transmission. Further, at the receiver, the incoming signal is normally filtered to minimize the negative effects of noise and interference. Filtering the signal results in changes to the shape of the original pulses, spreading their energy into adjacent pulses. This spreading effect is known as *dispersion* and can result in distortion of the pulse amplitude at the sampling instant unless carefully controlled (the sampling instant is the instant at which the receiver decides on the polarity, and hence the binary value, of the received pulse). In the *Nyquist criterion* on bandwidth transmission, it is shown [1] that the minimum real channel bandwidth that f_s independent symbols can be transmitted through, without resulting in symbol amplitude distortion at the sampling instant, is the Nyquist bandwidth $f_n = f_s/2$. Thus, for the rectangular pulse stream described above, the minimum transmission bandwidth is half the width of the main lobe. Because the stream is binary, one transmitted symbol contains one information bit. Thus the bit rate f_b of the stream is the same as the symbol rate f_s and the minimum real transmission bandwidth f_n is $f_b/2$.

In a *pulse amplitude modulation* (PAM) digital baseband system, bits are transmitted as symbols. If the transmitted stream is a two-level one, then the symbols are the bits. If the transmitted stream is a 2^n level one, where $n > 2$, however, incoming bits are converted to symbols. Let us consider the case of a four-level pulse amplitude modulation (PAM) stream. Here two information bits are encoded into each symbol, the four symbol levels being encoded to represent 00, 01, 10, and 11. As a result, the symbol rate f_s is $f_b/2$. In general, for L-level transmission systems, each transmitted symbol contains n information bits, where n is given by $n = \log_2 L$. Thus $f_s = f_b/n$, and the minimum transmission bandwidth $f_n = f_b/2n$. For example, for an eight-level PAM signal, $n = 3$ and thus $f_n = f_b/6$.

5.2.1   *Raised Cosine Filtering*

We will initially look at the transmission of impulses and then at the more practical case of pulse transmission. For impulse transmission, an ideal low pass brick-wall filter, of bandwidth $f_s/2$ and fixed time delay, will result in a nondistorted pulse amplitude at the sampling instant. Unfortunately, such a filter is unrealizable. Nyquist has shown, however, that if the amplitude characteristic of the brick-wall filter is modified to have a gradual roll-off, with odd symmetry about the Nyquist bandwidth f_n, then a nondistorted pulse amplitude at the sampling instant is preserved. One class of such a filter that is most often used in digital communication systems is the so-called raised cosine filter. The amplitude characteristic of such a filter for impulse transmission, $H_{im}(f)$, consists of a flat portion followed by a roll-off portion that has a sinusoidal form. It is characterized in terms of its *roll-off factor* α, where α is defined as f_x/f_n, f_x being the amount of bandwidth used in excess of the Nyquist bandwidth. The roll-off factor α may vary between 0 and 1, corresponding to an excess bandwidth of 0 to 100%. The amplitude characteristic $H_{im}(f)$ is given mathematically by

$$H_{im}\left(f\right) = 1, \quad 0 < f < f_n - f_x$$
$$= \frac{1}{2}\left[1 - \sin\frac{\pi}{2\alpha}\left(\frac{f}{f_n} - 1\right)\right], \quad f_n - f_x < f < f_n + f_x \tag{5.1}$$
$$= 0, \quad f > f_n + f_x$$

where $\alpha = f_x / f_n$

Figure 5.2, which is from Feher [1], is a graphical representation of $H_{im}(f)$.

The phase characteristic $\varphi(f)$ of the raised cosine filter is linear over the frequency range where the amplitude response is greater than zero and is thus given by

$$\phi\left(f\right) = Kf \quad 0 < f < f_n + f_x \tag{5.2}$$

where K is a constant.

Because the input to the filter defined above is a stream of impulses, then the amplitude spectral density at the filter input is of constant amplitude and, as per Eq. (3.15), the amplitude spectral density at the filter output $S_{rc}(f)$ has the identical characteristic of the filter transfer function $H_{im}(f)$, i.e.,

$$S_{rc}\left(f\right) = H_{im}\left(f\right) \tag{5.3}$$

A subtle but important point to note is that it is this output spectral density and hence the received pulse shape that results in nondistorted pulse amplitudes at the sampling instants, not the filter transfer function per se.

In practical systems, pulses of finite duration, not impulses, are used for digital transmission. A commonly used pulse is the full-length rectangular pulse. We recall from Eq. (3.3) that the spectrum of such a pulse has a (sin x)/x form. For nondistorted pulse amplitudes at the sampling instant, we desire that the spectral density, and hence the pulse shape, at the filter output of a transmission system conveying such pulses be the same as that for the impulse case discussed above, namely, $S_{rc}(f)$. To achieve this thus requires that the transfer function of the low pass filter, $H_{rp}(f)$ say, be the filter transfer function for impulses modified by the factor $x/(\sin x)$. Thus $H_{rp}(f)$ is given by

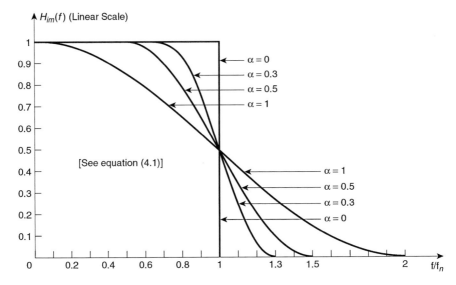

Fig. 5.2 Amplitude characteristics of the Nyquist channel for impulse transmission. (From [1], with the permission of the author)

$$H_{rp}\left(f\right) = \frac{\pi f \big/ 2f_n}{\sin\left(\pi f \big/ 2f_n\right)} H_{im}\left(f\right) \qquad (5.4)$$

Figure 5.3, which is also from Feher [1], is a graphical representation of $H_{rp}(f)$.

The general half-sided shape of time responses resulting from filtering to achieve a raised cosine spectral density $S_{rc}(f)$ as defined in Eq. (5.3) is shown in Fig. 5.4 for excess bandwidth values α of 0% and 50%. The full-sided time response extends from the negative to positive time axis with the pulse centered at time 0. Following pulses would be centered at time 1, 2, etc. The shape of these time responses, irrespective of roll-off factor, explains why there is no pulse amplitude distortion at the sampling instant as a consequence of the spreading of adjacent pulses. Pulses are sampled at their maximum amplitude. However, with a raised cosine spectrum, spread pulses adjacent to a pulse being sampled are always at zero amplitude at the time of sampling.

The essential components of a baseband digital transmission system are shown in Fig. 5.5. The input signal can be a binary or multilevel (>2) pulse stream. The transmitter low-pass filter, with transfer function $T(f)$, is used to limit the transmitted spectrum. Noise and other interference are picked up by the transmission medium and fed into the receiver filter. The receiver filter, with transfer function $R(f)$, minimizes the noise and interference relative to the desired signal. The output of the receiver filter is fed to a decision threshold unit which, for each pulse received, decides what was its most likely original level and outputs a pulse of this level. For a binary pulse stream, it outputs a pulse of amplitude +V say, if the

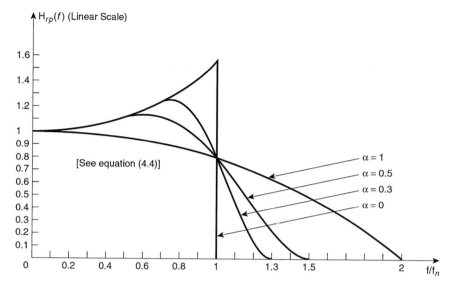

Fig. 5.3 Amplitude characteristics of the Nyquist channel for rectangular pulse transmission. (From [1], with the permission of the author)

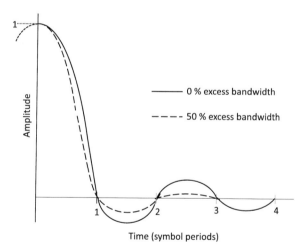

Fig. 5.4 Half-sided time responses resulting from raised cosine amplitude spectral density for values of excess bandwidth of 0% and 50%

input pulse is equal to or above its decision threshold, which is 0 volts. If the input pulse is below 0 volts, it outputs a pulse of amplitude $-V$.

In designing the system, a nondistorted pulse amplitude at the sampling instant is desirable at the output of the receive filter. This is normally achieved by employing filtering that results in a raised cosine amplitude spectral density S_{rc} at the receiver filter output. Thus, for full-length rectangular pulses, the combined transfer function of the transmitter filter and receiver filter (assuming that the transmission medium results in negligible impairment) should be as given in Eq. (5.4). There are an infinite number of ways of partitioning the total filtering transfer function between the transmitter filter and the receiver filter. Normally, the receiver filter is

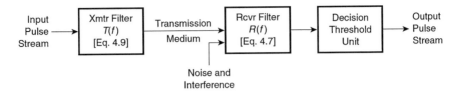

Fig. 5.5   Basic baseband digital transmission system

chosen to maximize the signal-to-noise ratio at its output as this optimizes the error
rate performance in the presence of noise. It can be shown [2, 3] that, for white
Gaussian noise, the receiver filter transfer function $R(f)$ that accomplishes this is
given by

$$R(f) = \left|S_{rc}(f)\right|^{\frac{1}{2}} \tag{5.5}$$

A filter with such a transfer function is referred to as a *root-raised cosine* (RRC)
filter. The transmitter filter transfer function $T(f)$ is then chosen to maintain the
desired composite characteristic, i.e.,

$$T(f).R(f) = H_{rp}(f) \tag{5.6}$$

Thus, by Eqs. (5.4), (5.5), and (5.6)

$$T(f).\left|S_{rc}(f)\right|^{\frac{1}{2}} = \frac{\pi f / 2f_n}{\sin\left(\pi f / 2f_n\right)} S_{rc}(f)$$

and hence

$$T(f) = \frac{\pi f / 2f_n}{\sin\left(\pi f / 2f_n\right)} \left|S_{rc}(f)\right|^{\frac{1}{2}} \tag{5.7}$$

Figure 5.6 shows plots of $R(f)$ and $T(f)$ for $\alpha = 0.5$

5.2.2   Integrate and Dump Filtering

A matched filter is the term used to define a transmit/receive filter combination that
gives the maximum signal-to-noise power ratio at its output for a given transmitted
symbol waveform in the presence of white Gaussian noise. Properly partitioned

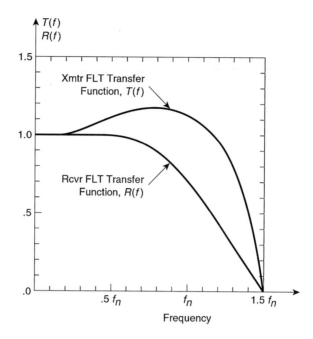

Fig. 5.6 Amplitude transfer functions of transmitter and receiver filters when α = 0.5

baseband filtering, as outline in Sect. 5.2.1 above, results in an optimum signal-to-noise power ratio at the receiver filter output. Such filtering is therefore matched filtering. This form of filtering is not, however, the only way to achieve matched filtering. For the situation where the transmitted baseband signal is a sequence of opposite polarity symbols $s_1(t)$ and $s_2(t)$ say, and where the symbol shapes are identical and hence $s_2(t) = -s_1(t)$, baseband signal recovery in the receiver can be achieved via an *integrate and dump* detector, and such a detector provides matched filtering [4].

An integrate and dump detector can be modelled as shown in Fig. 5.7a below. In such a detector, the received data stream data $s(t)$, consists of symbols $s_1(t)$ and $s_2(t)$, of identical form except of reversed polarity, both of length T, accompanied by noise. Stream $s(t)$ is first multiplied by a locally generated version of $s_1(t) - s_2(t)$ to create the signal stream $y(t)$. Each symbol of $y(t)$ then fed to the integrator where it is integrated over the symbol period T. At the end of the integration period, the integrator output $z(t)$ is sampled (or "dumped") and fed to a threshold device. What we get at the end of each sampling period is a signal with minimum probability of error. Immediately after each integration, all energy-storing elements in the integrator are discharged in preparation for the integration of the next symbol. The output of the threshold device, $\hat{s}(t)$, is the receiver's estimate of $s(t)$. If the integrate and dump output is ≥0, the threshold device outputs $s_1(t)$, if <0, it outputs $s_2(t)$. At the transmitter the original baseband stream is often a sequence of rectangular pulses. If these pulses are transmitted unfiltered, then $s_1(t)$ and $s_2(t)$ are of constant value over the symbol period T. The multiplier signal at the receiver is therefore a constant and can be omitted. Should the signal at the transmitter be filtered, however, $s_1(t)$ and $s_2(t)$

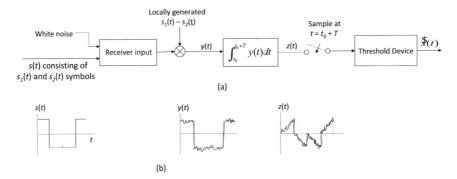

(a)

(b)

Fig. 5.7 Integrate and dump filtering. (**a**) Integrate and dump detector. (**b**) Data stream at various stages of integrate and dump process

will be of varying value with time. The multiplication process will then be required at the receiver. Notice that here "filtering" is defined as time domain shaping of the baseband symbols as opposed to frequency domain shaping in the case of raised cosine filtering. Figure 5.7b shows the processing of a received signal stream for the case where the transmitted symbols are unfiltered rectangular pulses. The integrator integrates the signal voltage such that its value increases linearly with time. It also integrates the noise. However, the noise voltage has zero mean, taking on positive and negative values. Its integrated value thus increases more slowly compared to the signal. This results in optimum signal-to-noise voltage at the end of the sampling period and hence optimum error performance.

Orthogonal frequency division multiplexing (OFDM), covered in Chap. 7, is a technique used in all 5G systems. OFDM is a transmission technique where several linearly modulated subcarriers are frequency division multiplexed. The transmitted symbols of an OFDM-modulated subcarrier are typically unfiltered and thus of constant amplitude. At the OFDM receiver, optimum detection is achieved via the integrate and dump process, hence our interest in this detection approach.

5.3   Linear Modulation Systems

Above we discussed PAM baseband systems. Wireless communication systems, however, operate in assigned frequencies that are considerably higher than baseband frequencies. It is thus necessary to employ modulation techniques that shift the baseband data up to the operating frequency. In this section we consider linear modulation systems.

These systems are so-called because they exhibit a linear relationship between the baseband signal and the modulated RF carrier. As a result of this relationship, their performance in the presence of noise and other impairments can be deduced from their equivalent baseband forms, hence our earlier review of baseband systems. We will commence this study by reviewing so-called *double-sideband suppressed carrier* (*DSBSC*) modulation as this modulation forms the foundation on which many of the most widely used linear modulation methods are based.

5.3.1   Double-Sideband Suppressed Carrier (DSBSC) Modulation

A simplified DSBSC system for PAM signal transmission is shown in Fig. 5.8. First a polar L-level PAM input signal, $a(t)$, with equiprobable symbols, is filtered with the low-pass filter, F_T, to limit its bandwidth to f_m say, and the filtered signal $b(t)$ applied to a multiplier. Also feeding the multiplier is a sinusoidal signal at the desired carrier frequency, f_c. As a result, the output signal of the multiplier, $c(t)$, is given by

$$c(t) = b(t)\cos 2\pi f_c t \tag{5.8}$$

If the amplitude spectral densities of $b(t)$ and $c(t)$ are represented as $B(f)$ and $C(f)$ respectively, then, by Eq. (3.9), $C(f)$ is given by

$$C(f) = \frac{1}{2}B(f + f_c) + \frac{1}{2}B(f - f_c) \tag{5.9}$$

Thus, as shown in Fig. 5.9, $C(f)$ consists of two spectra. One is real centered at f_c and the other imaginary centered at $-f_c$, and each has a bandwidth $2f_m$ and an amplitude half that of $B(f)$. As these spectra are symmetrically disposed on either side of the carrier frequency, the signal is referred to as a *double-sideband (DSB)* signal. Further, as $b(t)$ is polar, and thus has no fixed component, $c(t)$ contains no discrete carrier frequency component and is thus referred to as a *suppressed carrier* signal.

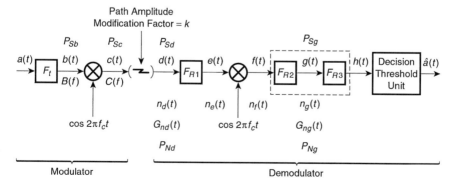

Fig. 5.8 Simplified one-way DSBSC system for PAM transmission

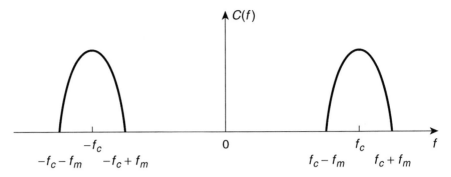

Fig. 5.9 DSBSC signal amplitude spectral density representation

We assume that $c(t)$ travels over a linear transmission path and arrives at the demodulator input modified in amplitude by the factor k. Thus, the input signal $d(t)$ to the receiver is given by

$$d(t) = k \cdot c(t) \tag{5.10}$$

The received signal $d(t)$ is passed through the bandpass filter F_{R1} to limit noise and interference. The bandwidth W of F_{R1} is normally greater than $2f_m$, the bandwidth of $d(t)$, so as to not impact the spectral density of $d(t)$. Assuming this to be the case, the output signal $e(t)$ of F_{R1} is given by

$$e(t) = d(t) \tag{5.11}$$

The signal $e(t)$ is fed to a multiplier. Also feeding the multiplier is the sinusoidal signal $\cos 2\pi f_c t$. As a result, the output of the multiplier $f(t)$ is given by

$$f(t) = e(t)\cos 2\pi f_c t \tag{5.12}$$

Substituting Eqs. (5.11), (5.10), and (5.8) into Eq. (5.12), we get

$$
\begin{aligned}
f(t) &= k \cdot b(t)\cos^2\left(2\pi f_c t\right) \\
&= \frac{k}{2} b(t)\left[1 + \cos\left(2 \cdot 2\pi f_c t\right)\right] \\
&= \frac{k}{2} b(t) + \frac{k}{2} b(t)\cos\left(2 \cdot 2\pi f_c t\right)
\end{aligned}
\tag{5.13}
$$

Thus, by multiplying $e(t)$ by $\cos 2\pi f_c t$, a process referred to as *coherent detection*, we recover $b(t)$ and create a second signal with the same double-sided bandwidth as $b(t)$ but centered at $2f_c$. The signal $f(t)$ is fed into the low-pass filter F_{R2} that eliminates the component of the signal centered about $2f_c$ while leaving the baseband component undisturbed. Thus, the output of F_{R2}, $g(t)$, is given by

$$g(t) = \frac{k}{2}b(t) \tag{5.14}$$

The signal $g(t)$ is fed to F_{R3} for final pulse shaping prior to level detection in the decision threshold unit. In practice F_{R2} and F_{R3} are combined into one but are shown separately here to add clarity to the analysis. The output, $\hat{a}(t)$, of the decision threshold unit is a PAM signal that is the demodulator's best estimate of the modulator input signal, $a(t)$.

5.3.2 Binary Phase Shift Keying (BPSK)

A special case of PAM transmission via a DSBSC system is when the PAM signal $a(t)$ in Fig. 5.8 has a binary, polar format. In this situation, if the filtered signal $b(t)$ has maximum peak amplitude of $\pm b$ volts say, then the modulated signal $c(t)$ varies between $c_1(t)$ and $c_0(t)$ as $b(t)$ varies between $+b$ and $-b$, where

$$c_1(t) = b\cos 2\pi f_c t \tag{5.15}$$

$$\begin{aligned} c_0(t) &= -b\cos 2\pi f_c(t) \\ &= b\cos(2\pi f_c t + \pi) \end{aligned} \tag{5.16}$$

When $b(t)$ is positive, the phase of $c(t)$ relative to the carrier phase is $0°$. When $b(t)$ is negative, the phase of $c(t)$ relative to the carrier phase is π radians or $-180°$. Thus, the relative phase has only two states. This modulation is referred to as *Binary Phase Shift Keying (BPSK)* and represents the simplest linear modulation scheme. Figure 5.10 shows typical examples of signals $a(t)$, $b(t)$, and $c(t)$. Figure 5.11 shows the *signal space* or *vector* or *constellation diagram* of $c(t)$. This diagram portrays both the amplitude and phase of $c(t)$ at the instances when the modulating signal $b(t)$ is at its peak.

It can be shown [5] that the probability of bit error $P_{be(BPSK)}$ of a BPSK system, with optimum filtering, and in the presence of white Gaussian noise, is given by

$$P_{be(BPSK)} = Q\left[\left(\frac{2P_S}{P_N}\right)^{\frac{1}{2}}\right] \tag{5.17}$$

where P_S = the average signal power at the demodulator input and P_N = the noise power in the two-sided Nyquist bandwidth at the demodulator input.

In Eq. (5.17) above, probability of error is defined in terms of a signal-to-noise ratio. In digital communication systems, however, it is just as common to define the probability of bit error, P_{be}, in terms of the ratio of the energy per bit E_b in the received signal to the noise power density N_0 at the receiver input. Defining P_{be} in

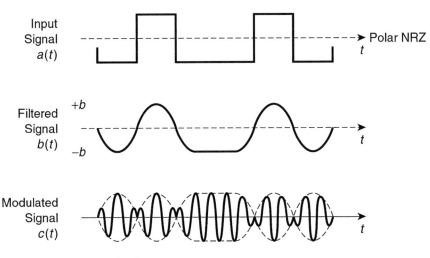

Fig. 5.10 Typical BPSK signals

Fig. 5.11 Signal space diagram of BPSK modulated signal

terms of E_b/N_0 makes it easy to compare the error performance of different modulation systems for the same bit rate. Given that, for a BPSK system with bit rate f_b, and hence a single-sided, double sideband Nyquist bandwidth also of f_b:

$$P_S = E_b \cdot f_b \tag{5.18}$$

and

$$P_N = \frac{N_0}{2} \cdot f_b \cdot 2 = N_0 f_b \tag{5.19}$$

Thus

$$\frac{P_S}{P_N} = \frac{E_b}{N_0} \tag{5.20}$$

and hence

$$P_{be(BPSK)} = Q\left[\left(2\frac{E_b}{N_0}\right)^{\frac{1}{2}}\right] \tag{5.21}$$

Figure 5.12 shows the power spectral density of BPSK when the modulating signal is unfiltered. This power spectral density has the same $\sin x/x$ form as that of the two-sided baseband signal, except that it is shifted in frequency by f_c. It can be shown [4] to be given by

$$G_{BPSK}(f) = P_S \tau_b \left[\frac{\sin \pi (f - f_c)\tau_b}{\pi (f - f_c)\tau_b}\right]^2 \tag{5.22}$$

where τ_b, which is equal to $1/f_b$, is the bit duration of the baseband signal.

Also shown in Fig. 5.12 is the single-sided, double sideband Nyquist bandwidth of the system, which is equal to f_b. Thus, at its theoretical best, BPSK is capable of transmitting 1 bit per second in each Hertz of transmission bandwidth. The system is therefore said to have a maximum *spectral efficiency* of 1 bit/s/Hz.

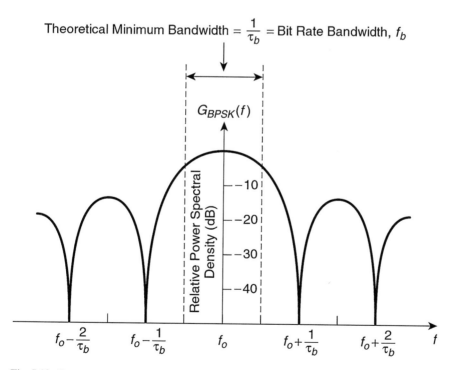

Fig. 5.12 Power spectral density of BPSK

Because, as indicated earlier, filtering to 0% excess bandwidth to achieve the Nyquist bandwidth is not practical, real BPSK systems have spectral efficiencies less than 1 bit/s/Hz. For an excess bandwidth of 25% say, then data at a rate of 1 bit/sec requires 1.25 Hertz of bandwidth, leading to spectral efficiency of 0.8 bits/s/Hz.

As we shall see in succeeding sections, spectral efficiencies much greater than that afforded by BPSK are easily realizable. As a result, BPSK is rarely used in wireless communication networks, where, as a rule, available spectrum is limited and thus highly valued. Nonetheless, an understanding of its operating principles is very valuable in analyzing *Quadrature Phase Shift Keying* (QPSK), a popular modulation technique.

For BPSK modulated with a pulse train of rectangular pulses, for example, signal $a(t)$ shown in Fig. 5.10, the amplitude of the modulated signal is unchanged as the pulse train progresses, as a change in pulse polarity simply flips the phase but leaves the amplitude unchanged. Thus, since power is proportional to signal level squared, the peak-to-average power ratio (PAPR) of this modulated signal is 1. When the modulating signal is filtered, however, for example, signal $b(t)$ shown in Fig. 5.10, the amplitude of the modulated signal changes with changes in the polarity of the modulating signal, decreasing from maximum to a minimum of 0. The average signal power is thus less than the peak signal power. For limited filtering the PAPR is likely to be slightly above 1 but for significant filtering is more likely to be in the region of 2.

5.3.3 Pi/2 BPSK

Pi/2 BPSK is simply BPSK with a $\pi/2$ counterclockwise phase shift rotation of the carrier frequency on every successive symbol. On the constellation diagram, BPSK occupies two phase positions, 0 and π radians, whereas Pi/2 BPSK occupies four phase positions, namely, o, $\pi/2$, π, and 3/2 π radians. To visualize phase transitions with Pi/2 BPSK, take, for example, the case where the current phase position is 0 radians. For the next symbol, there is an automatic $\pi/2$ phase rotation. Thus, if that next symbol were to be of value binary 1, the next phase location would be $\pi/2$, but were it to be of value binary 0, the next phase location would be 3/2 π. Pi/2 BPSK exhibits the same bit error rate performance and the same spectral density characteristics as BPSK.

With Pi/2 BPSK, constellation phase changes are limited to $\pi/2$ radians. Like BPSK, the PAPR of Pi/2 BPSK is 1 if the modulated symbols are rectangular as the modulated signal amplitude never changes. However, if the modulating symbols are filtered, the modulated signal amplitude, unlike with BPSK, never goes through 0. In fact, it can be easily shown that the ratio of its maximum amplitude relative to its minimum is $\sqrt{2}$. This leads to a PAPR much less than that of BPSK for the same level of filtering.

5.3.4   Quadrature Amplitude Modulation (QAM)

The BPSK system described above is only capable of amplitude modulation accompanied by 0° or 180° phase shifts. However, by adding a *quadrature branch* as shown if Fig. 5.13, it becomes possible to generate signals with any desired amplitude and phase. In the quadrature branch, a second PAM baseband signal is multiplied with a sinusoidal carrier of frequency f_c, identical to that of the *in-phase* carrier but delayed in phase by 90°. The outputs of the two multipliers are then added together to form a *Quadrature Amplitude Modulated* (QAM) signal.

Labeling the in-phase PAM filtered signal $b_i(t)$ and the quadrature PAM filtered signal $b_q(t)$, then the summed output of the modulator, $c(t)$, is given by

$$c(t) = b_i(t)\cos 2\pi f_c t + b_q(t)\sin 2\pi f_c t \qquad (5.23)$$

At the QAM demodulator, the incoming signal is passed through the bandpass filter F_{R1} to limit noise and interference. It is then divided into two and each branch inputted to a multiplier, one multiplier being fed also with an in-phase carrier, $\cos 2\pi f_c t$, and the other with a quadrature carrier, $\sin 2\pi f_c t$.

The output signal $f_i(t)$ of the in-phase multiplier is given by

$$
\begin{aligned}
f_i(t) &= k \cdot c(t) \cdot \cos 2\pi f_c t \\
&= k \cdot b_i(t)\cos^2 2\pi f_c t + k \cdot b_q(t)\sin 2\pi f_c(t) \cdot \cos 2\pi f_c t \qquad (5.24)\\
&= \frac{k}{2} b_i(t) + \frac{k}{2} b_i(t)\cos(2 \cdot 2\pi f_c t) + \frac{k}{2} b_q(t)\sin(2 \cdot 2\pi f_c t)
\end{aligned}
$$

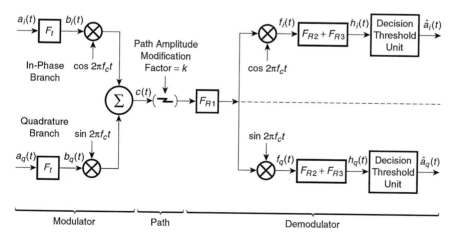

Fig. 5.13 Simplified one-way quadrature amplitude modulated system

The only difference between $f_i(t)$ and $f(t)$ of Eq. (5.13) for an in-phase only modulated system is the final component of Eq. (5.24). However, this component, like the second component in the Eq. (5.24), is spectrally centered at $2f_c$ and is filtered prior to decision threshold detection, leaving only the original quadrature modulating signal $b_i(t)$.

The output signal $f_q(t)$ of the quadrature multiplier is given by

$$f_q(t) = k \cdot c(t) \cdot \sin 2\pi f_c t$$
$$= k \cdot b_q(t) \sin^2 2\pi f_c t + k \cdot b_i(t) \cos 2\pi f_c t \cdot \sin 2\pi f_c t \qquad (5.25)$$
$$= \frac{k}{2} b_q(t) - \frac{k}{2} b_q(t) \cos(2 \cdot 2\pi f_c t) + \frac{k}{2} b_i(t) \sin(2.2\pi f_c t)$$

As with the output $f_i(t)$ from the in-phase multiplier, $f_q(t)$ consists of the original in-phase modulating signal $b_q(t)$ as well as two components centered spectrally at $2f_c$ which are filtered prior to decision threshold unit. Thus, by quadrature modulation, it is possible to transmit two independent bit streams on the same carrier with no interference of one signal with the other, given ideal conditions.

5.3.5 Quadrature Phase Shift Keying (QPSK)

Quadrature (or Quaternary) Phase Shift Keying (QPSK) is one of the simplest implementations of Quadrature Amplitude Modulation and is sometimes referred to as *4-QAM*. In it, the modulated signal has four distinct states. A block diagram of a conventional, simplified, QPSK system is shown in Fig. 5.14a. The binary non-return-to-zero (NRZ) input data stream $a_{in}(t)$, of bit rate f_b and bit duration τ_b, is fed to the modulator where it is converted by a serial to parallel converter into two NRZ streams, an *I* stream labeled $a_i(t)$ and a *Q* stream labeled $a_q(t)$, each of symbol rate f_B, half that of f_b, and symbol duration τ_B, twice that of τ_b. The relationship between the data streams $a_{in}(t)$, $a_i(t)$, and $b_q(t)$ is shown in Fig. 5.14b. The *I* and *Q* streams undergo standard QAM processing as described in Sect. 5.3.4 above. The in-phase multiplier is fed by the carrier signal cos $2\pi f_c t$. The quadrature multiplier is fed by the carrier signal delayed by 90° to create the signal sin $2\pi f_c t$. The output of each multiplier is a BPSK signal. The BPSK output signal of the in-phase carrier-driven multiplier has phase values of 0° and 180° relative to the in-phase carrier, and the BPSK output signal of the quadrature carrier-driven multiplier has phase values of 90° and 270° relative to the in-phase carrier. The multiplier outputs are summed to give a four-phase signal. Thus, QPSK can be regarded as two *associated BPSK* (ABPSK) systems operating in quadrature.

The four possible output signal states of the modulator, their *IQ* digit combinations, and their possible transitions from one state to another are shown in Fig. 5.14c. We note that either 90° or 180° phase transitions are possible. As an example, a 90° phase transition occurs when the *IQ* combination changes from 00 to 10, and a 180°

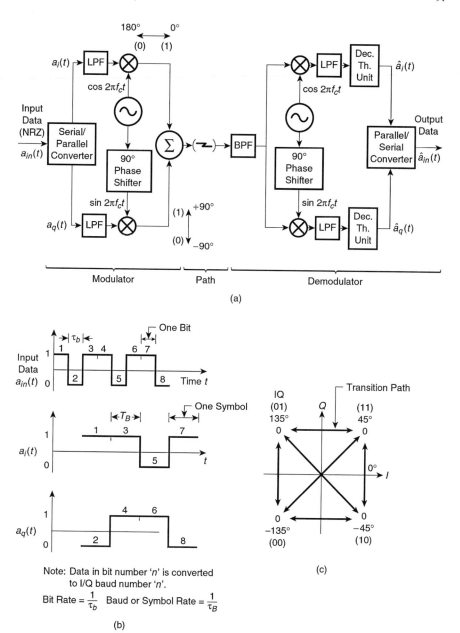

Fig. 5.14 QPSK system representation. (**a**) Block diagram. (**b**) Modular data streams. (**c**) Signal space diagram

phase transition occurs when the *IQ* combination changes from 00 to 11. For a system where $a_i(t)$ and $a_q(t)$ are unfiltered prior to application to the multipliers, phase transitions occur instantaneously, and thus the signal has a constant amplitude. However, for systems where these signals are filtered, as is normally the case to limit the radiated spectrum, phase transitions occur over time, and the modulated signal has an amplitude envelope that varies with time. In particular, a 180° phase change results in a change over time in amplitude envelope value from maximum to zero and back to maximum.

In the demodulator, as a result of quadrature demodulation, signals $\widehat{a}_i(t)$ and $\widehat{a}_q(t)$, estimates of the original modulating signals are produced. These signals are then recombined in a parallel to serial converter to form $\widehat{a}_{in}(t)$, an estimate of the original input signal to the modulator.

As indicated above, QPSK can be regarded as two associated BPSK systems operating in quadrature. From a spectral point of view at the modulator output, two BPSK signal spectra are superimposed on each other. The BPSK symbol duration is τ_B. But $\tau_B = 2\tau_b$. Thus the spectral density of each BPSK signal, and hence of the QPSK signal, is given by Eq. (5.22), but with τ_b replaced by $2\tau_b$. Making this replacement, we get

$$G_{QPSK}(f) = 2P_s\tau_b \left[\frac{\sin 2\pi (f - f_c)\tau_b}{2\pi (f - f_c)\tau_b} \right] \tag{5.26}$$

A graph of $G_{QPSK}(f)$ is shown in Fig. 5.15. We note that the widths of the main lobe and side lobes are half that for BPSK given the same bit rate for each system. As a result, the maximum spectral efficiency of QPSK is twice that of BPSK, i.e., 2 bits/s/Hz.

It can be shown [5] that the probability of bit error $P_{be(QPSK)}$ of a QPSK system, with optimum filtering, and in the presence of white Gaussian noise, is given by

$$P_{be(QPSK)} = Q\left[\left(\frac{2E_b}{N_0} \right)^{\frac{1}{2}} \right] \tag{5.27}$$

We note that this relationship is identical to that for the probability of bit error versus E_b/N_0 for BPSK. Graphs of P_{be} versus E_b/N_0 for QPSK and other linear modulation methods are shown in Fig. 5.16.

In summary, for the same bit rate, the spectral efficiency of QPSK is twice that of BPSK with no loss in the probability of bit error performance in ideal circumstances. The QPSK hardware is, however, more complex than that required for BPSK. Further, in transmission through nonlinear components such as power amplifiers, filtered QPSK is subject to quadrature crosstalk, a situation where modulation on one quadrature channel ends up on the other. This situation can also arise in the receiver if the phase difference between the coherent detection

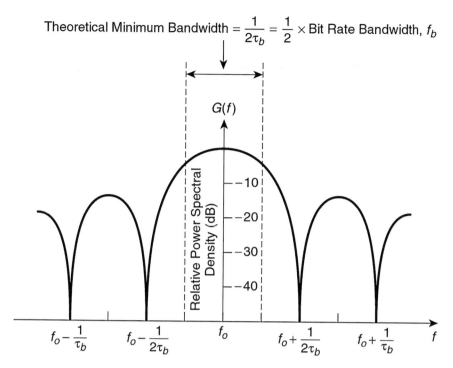

Fig. 5.15 Power spectral density of QPSK

oscillators is not kept to 90°. Thus, in real world environments, BPSK is a more robust modulation scheme than QPSK.

5.3.6   High-Order 2^{2n}-QAM

Though relatively easy to implement and robust in performance, linear four phase systems such as QPSK do not often afford the desired spectral efficiency in commercial wireless systems. Higher-order QAM systems, however, do permit higher spectral efficiencies and have become very popular. A common class of QAM systems allowing high spectral density is one where the number of states is 2^{2n}, where n equals 2, 3, 4, ... A generalized and simplified block diagram of a 2^{2n}-QAM system is shown in Fig. 5.17. The difference between this generalized system and the QPSK system shown in Fig. 5.14 is that (a) in the generalized modulator, the I and Q signals $a_i(t)$ and $a_q(t)$ are each fed to a 2 to 2^n level converter prior to filtering and multiplication with the carrier and (b) in the generalized demodulator, the outputs of the decision threshold units are each fed to a 2^n to 2 level converter prior to being combined in a parallel to serial converter. 2^{2n}-QAM systems have been deployed commercially for values of n from 2 to 7.

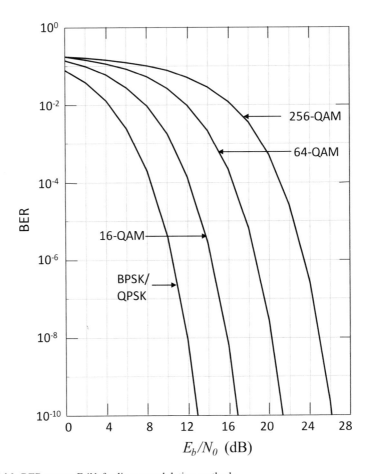

Fig. 5.16   BER versus E_b/N_0 for linear modulation methods

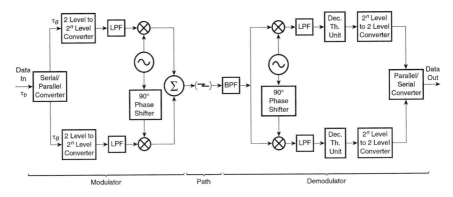

Fig. 5.17   Generalized block diagram of 2^{2n}-QAM system

For *n* equal 2, a 16-QAM system is derived. In such a system, incoming symbols to each modulator level converter are paired, and output symbols, in the form of signals at one of four possible amplitude levels, are generated in accordance with the coding table shown in Fig. 5.18a. The duration of these output symbols, τ_{B4L} say, is twice that of, τ_B, the duration of the input symbols. As a result of the application of the 4-level signals to the multipliers, the output of each multiplier is a 4-level amplitude modulated DSBSC signal, and the combined signal at the modulator output is a QAM signal with 16 states. Thus, 16-QAM can be treated as two 4-level PAM DSBSC systems operating in quadrature. The constellation diagram of a 16-QAM signal is shown in Fig. 5.18b. From this figure it is clear that 16-QAM has an amplitude envelope that varies considerably over time, irrespective of whether the signal has been filtered or not, and thus must be transmitted over a highly linear system if it is to preserve its spectral properties.

As τ_B, the duration of symbols from the serial to parallel converter, is equal to $2\tau_b$, where τ_b is the duration of incoming bits to the modulator, it follows that

$$\tau_{B4L} = 4\tau_b \tag{5.28}$$

Using Eq. (5.28) and the same logic used to determine the spectral density of QPSK, it can be shown that $G_{16-QAM}(f)$, the spectral density of 16-QAM, is given by

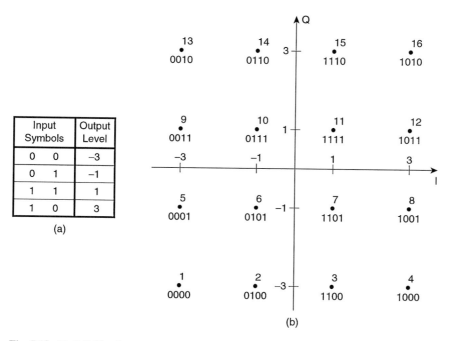

Fig. 5.18 16-QAM level converter coding table and constellation diagram. (**a**) 2-level to 4-level coding table. (**b**) Constellation diagram

$$G_{16-QAM}(f) = 4P_s\tau_b \left[\frac{\sin 4\pi (f-f_c)\tau_b}{4\pi (f-f_c)\tau_b} \right] \tag{5.29}$$

$G_{16-QAM}(f)$ is such that its main lobes and side lobes are one fourth as wide as those of BPSK. As a result, the maximum spectral efficiency off 16-QAM is 4 bits/s/Hz, twice that of QPSK.

For a 16-QAM system, it can be shown [5] that the probability of bit error $P_{be(16-QAM)}$ is given by

$$P_{be(16-QAM)} = \frac{3}{4}Q\left[\left(\frac{4}{5}\frac{E_b}{N_0} \right)^{\frac{1}{2}} \right] \tag{5.30}$$

A graph of $P_{be(16-QAM)}$ versus E_b/N_0 is shown in Fig. 5.16. It will be observed that for a probability of bit error of 10^{-3}, the E_b/N_0 required for 16-QAM is 3.8 dB greater than that required for QPSK. Thus, the doubling of the spectral efficiency achieved by 16-QAM relative to QPSK comes at the expense of probability of bit error performance.

The two-level to four-level coding shown in Fig. 5.18a is an example of *Gray coding*. In Gray coding, the bits that create any pair of adjacent levels differ by only one bit. It is interesting to note that, as a result of Gray coding in the *I* and *Q* channels, the 16 states shown in the signal space diagram in Fig. 5.18b are also Gray coded. Thus, an error resulting from one of these states being decoded as one of its closest adjacent states will result in only one bit being in error.

For *n* equal 3, a 64-QAM system is derived, 2 8-level PAM DSBSC signals being combined in quadrature. The eight-level PAM signals are created by grouping the incoming symbols to the level converter into sets of three and using these three-digit code words to derive the eight output levels.

For *n* equal 4, a 256-QAM system is derived, 2 16-level PAM DSBSC signals being combined in quadrature. Here the 16-level PAM signals are created by grouping the incoming symbols to the level converter into sets of 4 and using these 4-digit code words to derive the 16 output levels.

Using the same logic as applied to the above analysis of 16-QAM, but with generalized equations, it can be shown that the maximum spectral efficiency of 2^{2n}-QAM is $2n$ bits/s/Hz. Thus, that of 64-QAM is 6 bits/s/Hz and that of 256-QAM is 8 bits/s/Hz.

It is shown in [5] that for Gray-coded 2^{2n}-QAM, with optimum filtering, and in the presence of white Gaussian noise, the generalized equation for probability of bit error P_e versus E_b/N_0 is

$$P_{be(2^{2n}-QAM)} = \frac{2}{\log_2 L}\left[1-\frac{1}{L}\right]Q\left[\left(\frac{6\log_2 L}{L^2-1}\frac{E_b}{N_0} \right)^{\frac{1}{2}} \right] \tag{5.31}$$

where $L = 2^n$.

Thus, for 64-QAM we have

$$P_{be(64-QAM)} = \frac{7}{12} Q \left[\left(\frac{2}{7} \frac{E_b}{N_0} \right)^{\frac{1}{2}} \right] \tag{5.32}$$

and for 256-QAM we have

$$P_{be(256-QAM)} = \frac{15}{32} Q \left[\left(\frac{8}{85} \frac{E_b}{N_0} \right)^{\frac{1}{2}} \right] \tag{5.33}$$

The probability of error relationships for 64 and 256-QAM are also shown in Fig. 5.16. We observe that as the number of QAM states increases, the P_e performance decreases.

In the QAM realizations discussed above, the I and Q carriers were modulated via first a parallel to series converter, followed by, for the cases where n was greater than 1, a two-level to 2^n level converter. We note, however, that though helpful in conveying the modulation conceptually, such a physical realization is not necessary. All that is necessary is to utilize any mapping structure that converts each grouping of 2^n incoming data bits to the desired I and Q modulating values. Thus, with 16-QAM, for example, the mapper needs only be programmed to take any incoming 4-bit combination and map it to the I and Q values shown in Fig. 5.18. For example, incoming bits 1110 are mapped to an I value of 1 and a Q value of 3. Viewed another way, it's mapped to the complex value $1 + j3$. As we shall see, in 5G NR modulation, such mapping is employed.

5.3.7 Peak-to-Average Power Ratio

For the unfiltered BPSK, QPSK, and QAM systems studied above, the signal level is constant during each entire symbol. Thus, for BPSK and QPSK, symbol level is constant throughout an entire symbol stream, and as a result, the peak level equals the average level and peak power equals average power. Therefore, the peak-to-average power ratio (PAPR) is 1 or 0 dB. For 2^{2n}-QAM systems, however, where $n > 1$, the symbols can take several levels, leading to a PAPR greater than 1. Consider the 16-QAM constellation shown in Fig. 5.19. Here we see that the symbols can take one of four different levels. The highest level is $\sqrt{18}x$ and thus peak power is $18x^2$. The average power, $\overline{X^2}$, is given by

$$\overline{X^2} = \frac{4 \times 2x^2 + 8 \times 10x^2 + 4 \times 18x^2}{16} = 10x^2 \tag{5.34}$$

Fig. 5.19 Symbol levels
on a 16-QAM constellation
diagram

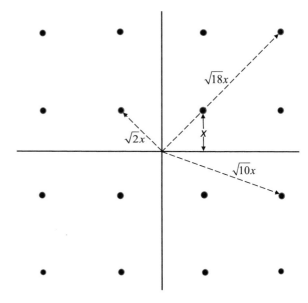

Thus, for unfiltered symbols, the PAPR equals $18x^2/10x^2 = 1.8 = 2.55$ dB. It can similarly be shown that, for unfiltered symbols, the PAPR of 64-QAM is 3.69 dB and that of 256-QAM is 4.23 dB.

5.4   Power Amplifier Linearization

The transmitter RF power amplifier follows the baseband to RF frequency shifting operation, and its purpose is to provide a high level of output power so that adequate signal level is available to the receiver even with significant fading. The output power of 5G transmitter power amplifiers varies from tenths of watts for mobile units to up to tens of watts for based stations, with maximum attainable power decreasing with frequency.

For linear modulation systems with signal states of varying amplitudes, linear amplification is essential to maintain acceptable performance, with higher and higher linearity required as the number of modulation states increases. The effects of nonlinearity on such systems are a nonlinear displacement of signaling states in the phase plane and the regeneration of spectral side lobes removed by prior filtering. The displacement of the signaling states degrades the error probability performance, while the regenerated spectrum can cause interference to signals in adjacent channels. To avoid these effects, the power amplifier must be capable of linearly amplifying all signaling states and thus amplifying the peak signal power. However, high-order linear modulation results in signals where the ratio between peak power and average power can be several dBs. Thus, power amplifiers processing these signals must operate at an average power that is backed off from the peak linear power available by a minimum of the peak to average ratio of the amplified signal.

The output signal of the RF power amplifier is normally fed via a bandpass filter to the antenna. The bandwidth of this filter is typically wider than the signal spectrum, and its purpose is to limit out of band radiation as well as to allow duplexing of this signal with an associated incoming signal.

In a number of wireless transmitter designs, the nonlinearity of the RF power amplifier is counteracted by employing a linearization technique such as *predistortion*. Predistortion works by purposely inserting a nonlinearity into the signal feeding the RF power amplifier that is the complement of the nonlinearity of the RF power amplifier. Many modern predistorters are digital and adaptive. In such designs, circuits are added that continuously measure the nonlinearity at the RF amplifiers output and feed this measurement back to the predistorter that adjusts its nonlinearity in such a way as to minimize the RF output nonlinearity. By using predistorters, it is typically possible to increase transmitter output power by low single digit dBs relative to non-predistorted output power.

5.5   The Receiver "Front End"

In receivers where the incoming RF signal is directly converted to baseband, as is normally the case with OFDM based systems (Sect. 7.2.1), the term *receiver front end* is used to cover combination of the receiver RF input bandpass filter, the *low noise amplifier* (LNA), and the downconverter. The purpose of the RF input filter is to eliminate out of band unwanted signals. The low-noise amplifier follows the input filter and plays a large part in determining the overall noise performance of the receiver. The characteristic of the receiver that determines the signal-to-noise ratio presented to the demodulator is the receiver *noise factor*. For the receiver shown in Fig. 5.20, the noise factor F_{RX} describes the deterioration of the signal-to-noise ratio from the input to the RF bandpass filter to the output of the downconverter and is given by:

$$F_{RX} = \frac{P_{Si} / P_{ni}}{P_{So} / P_{no}} \tag{5.35}$$

where P_{Si} = RF bandpass filter input signal power; P_{So} = Downconverter output signal power; P_{ni} = RF bandpass filter input noise power in a frequency band df; P_{no} = Downconverter output noise power in a frequency band df.

Thus

$$P_{So} / P_{no} \left(\text{dB} \right) = P_{Si} \left(\text{dBm} \right) - P_{ni} \left(\text{dBm} \right) - F_{RX} \left(\text{dB} \right) \tag{5.36}$$

where 1 dBm = 1 mW.

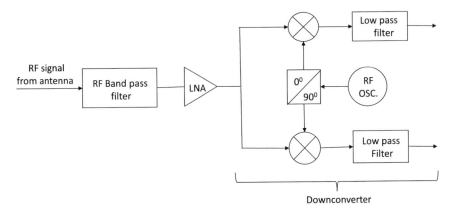

Fig. 5.20 Receiver front end

The signal-to-noise ratio at downconverter output, P_{So}/P_{no}, is also the signal-to-noise ratio at the demodulator input and thus that which largely determines the error rate performance. Equation (5.36) is important because it indicates that P_{So}/P_{no} can be determined from a knowledge of the signal-to-noise ratio at the receiver input and the receiver noise factor.

The thermal noise power in Watts available in a small frequency band df Hertz from a source having a noise temperature T degrees Kelvin is given by

$$P_n = k \cdot T \cdot df \tag{5.37}$$

where $k = 1.38 \cdot 10^{-23}$ Joules/degree Kelvin (Boltzmann's constant)

For a terrestrial wireless system, the source of thermal noise at the receiver input is the receiving antenna. Antenna noise temperature is normally assumed to be 290°Kelvin. At this temperature, the antenna noise transferred to the receiver is given by Eq. (5.37) to be −174 dBm per Hertz of bandwidth. Thus, the input thermal noise in the bit rate bandwidth f_b Hertz of a digital wireless system is given by

$$P_{nib}(\mathrm{dBm}) = -174 + 10\log_{10} f_b \tag{5.38}$$

Substituting Eq. (5.38) into Eq. (5.36) gives the ratio of the downconverter output signal power P_{So} to noise power in the bit rate bandwidth P_{nob} to be

$$P_{So} / P_{nob}(\mathrm{dB}) = P_{Si}(\mathrm{dBm}) + 174 - 10\log_{10} f_b - F_{RX}(\mathrm{dB}) \tag{5.39}$$

Recognizing that

$$\frac{P_{So}}{P_{nob}} = \frac{E_b f_b}{N_0 f_b} = \frac{E_b}{N_0} \tag{5.40}$$

where E_b is the energy per bit at the downconverter output/demodulator input and N_0 is the noise power spectral density at the downconverter output/demodulator input

Then Eq. (5.39) can be restated as

$$E_b / N_0 (\text{dB}) = P_{Si} (\text{dBm}) + 174 - 10 \log_{10} f_b - F_{RX} (\text{dB}) \qquad (5.41)$$

Thus, knowing the received input signal level, the bit rate, and the receiver noise factor, one can calculate the E_b/N_0 at the demodulator input and, from the appropriate probability of error versus E_b/N_0 relationship, the theoretical probability of error.

If the noise factor of the LNA is F_{LNA}, the gain of the LNA is G_{LNA}, and the noise factor at the input to the downconverter is F_{DC}, then by the well-known *Friis formula*, the noise factor at the input to the LNA, F_{LNA-IN}, is given by

$$F_{LNA-IN} = F_{LNA} + \frac{F_{DC} - 1}{G_{LNA}} \qquad (5.42)$$

This noise factor is not, however, the noise factor at the input to the receiver as the RF band pass filter and any other circuitry ahead of the LNA are likely to have a small but measurable loss. If this loss is L dB say, then the receiver input noise factor, expressed in dBs and hence referred to as the noise figure, is given by F_{RX}, where

$$F_{RX} = (L + 10 \log F_{LNA-IN}) \text{ dB} \qquad (5.43)$$

5.6   Summary

The fundamental modulation methods used in 5G are Pi/2 BPSK, QPSK, 16-QAM, 64-QAM, and 256-QAM. These methods were studied in this chapter, along with some issues involved in their realization. The mobile transmission path is a variable one. Thus, for user data transmission, in some situations, the SINR may be so high as to allow the use of 256-QAM. In other situations, the SINR may be so low as to require the use of QPSK. As the situation for an individual user can be constantly changing, the most efficient way to address this variability is to have "adaptive modulation," and this is indeed done in 5G NR. This action on its own, however, cannot provide the bit error rate reliability required of mobile broadband systems. Such reliability can only be achievable by the use of coding techniques along with adaptive modulation. In the following chapter, Chap. 6, channel coding methods applied in 5G NR are presented.

References

1. Feher K (1981) Digital communications: microwave applications. Prentice-Hall, Upper Saddle River
2. Lucky RW, Salz J, Weldon EJ (1968) Principles of data communication. McGraw-Hill, Inc, New York
3. Bennett WR, Davey JR (1965) Data transmission. McGraw-Hill, Inc, New York
4. Taub H, Schilling DL (1971) Principles of communication systems. McGraw-Hill, Inc, New York
5. Morais DH (2004) Fixed broadband wireless communications: principles and practical applications. Prentice-Hall, Upper Saddle River

Chapter 6
Channel Coding and Link Adaptation

6.1 Introduction

Coding, in the binary communications world, is the process of adding a bit or bits to useful data bits in such a fashion as to facilitate the detection or correction or errors incurred by such useful bits as a result of their transmission over a non-ideal channel. Such a channel, for example, may be one that adds noise, or interference, or unwanted nonlinearities. In this chapter we focus on error detection and error correction coding as applied in 5G NR, including the very important low-density parity check (LDPC) codes, polar codes, and hybrid automatic repeat request (HARQ). Error detection on its own obviously does nothing to improve error rate performance. However, as we shall see, it can aid in error correction when combined with other techniques.

Where helpful in understanding the various techniques presented, some basic coding theory and practice is presented. Several coding-related mathematical computations addressed below that involve binary data apply modulo 2 arithmetic. Thus, following is a review of *modulo 2 addition, modulo 2 subtraction, and modulo 2 multiplication:*

Addition	Subtraction	Multiplication
$0 + 0 = 0$	$0 - 0 = 0$	$0 \times 0 = 0$
$0 + 1 = 1$	$0 - 1 = 1$	$0 \times 1 = 0$
$1 + 0 = 1$	$1 - 0 = 1$	$1 \times 0 = 0$
$1 + 1 = 0$	$1 - 1 = 0$	$1 \times 1 = 1$

The original version of this chapter was revised. The correction to this chapter is available at
https://doi.org/10.1007/978-3-030-51441-9_12

© Springer Nature Switzerland AG 2020, Corrected Publication 2021   103
D. H. Morais, *Key 5G Physical Layer Technologies*,
https://doi.org/10.1007/978-3-030-51441-9_6

6.2 Error Detection: Cyclic Redundancy Check (CRC) Codes

Cyclic redundancy check (CRC) [1] is one of the most popular methods of error detection, being normally used with automatic repeat request (ARQ) (see Sect. 6.6 below) to facilitate error correction. It is employed by 5G NR. To create a CRC codeword, a mathematical calculation is carried out on a block of useful binary data which is referred to as a *Frame*. In the calculation, a string of n 0 s is appended to the Frame. This number is divided by another binary number derived from the *generator polynomial*. The number n is one less than the number of bits in the divisor. The remainder of this division represents content of the Frame and is added to the Frame. These added bits are called the *checksum*. At the receiver, the entire received codeword, i.e., the Frame plus the checksum, is divided by the same polynomial derived bits used by the transmitter. If the result of this division is zero, then the received Frame is assumed to be error free, if not, an error or errors is assumed to have occurred. Below is an example of CRC codeword generation.

Example 6.1 CRC Codeword Generation Let frame be 1 1 0 0 1.

Let the generator polynomial be of second-degree and equal $x^2 + 1$, leading to the binary number equivalent of $100 + 1 = 101$. Then number of bits in divisor is 3, thus number of zeros to be added to frame prior to division is 2. Subtraction in the division is modulo 2 subtraction. Division is therefore:

$$
\begin{array}{r}
101 \overline{)1\,1\,0\,0\,1\,0\,0} \\
1\,0\,1 \\
\hline
1\,1\,0 \\
1\,0\,1 \\
\hline
1\,1\,1 \\
1\,0\,1 \\
\hline
1\,0\,0 \\
1\,0\,1 \\
\hline
0\,1\,0 \\
0\,0\,0 \\
\hline
1\,0 \text{-} CRC
\end{array}
$$

If leftmost remainder digit is 0 then must use 0 0 0 → 0 0 0

Thus, generated codeword is 1 1 0 0 1 1 0.

CRC codes can detect all burst errors that affect an odd number of bits. They can detect all burst errors of length less than or equal to the degree of the polynomial. Further, they can detect, with a very high probability, burst errors of length greater than the degree of the polynomial.

In NR, the size of the CRC is a function of the size of the transport block, i.e., the size of the block of data being processed. For transport blocks larger than 3824 bits, a 24-bit CRC is added. For transport blocks less than or equal to 3824 bits, a 16-bit CRC is used to reduce overhead.

6.3   Forward Error Correction Codes

6.3.1   Introduction

Error control coding is a means of permitting the robust transmission of data by the deliberate introduction of redundancies into the data creating a codeword. One method of accomplishing this is to have a system that looks for errors at the receive end and, once an error is detected, makes a request to the transmitter for a repeat transmission of the codeword. In this method, called ARQ and alluded to above, a return path is necessary. Error correction coding that is not reliant on a return path inherently adds less delay to transmission. Such coding is referred to as *forward error correction* (FEC) coding. For digitally modulated signals, detected in the presence of noise, use of FEC results in the reduction of the residual BER, usually by several orders of magnitude and a reduction of the receiver 10^{-6} threshold level by about one to several dBs depending on the specific scheme employed. Figure 6.1 shows typical error performance characteristics of an uncoded versus FEC coded digitally modulated system. The advantage provided by a coded system can be quantified by *coding gain*. The coding gain provided by a particular scheme is defined as the reduction in E_b/N_0 in the coded system compared to the same system but uncoded for a given BER and the same data rate. Coding gain varies significantly with BER, as can be seen from Fig. 6.1, and above a very high level may even be negative. As BER decreases, the coding gain increases until it approaches a limit as the BER approaches zero (zero errors). This upper limit is referred to as the *asymptotic coding gain* of the coding scheme.

FEC works by adding extra bits to the bit stream prior to modulation according to specific algorithms. These extra bits contribute no new message information. However, they allow the decoder, which follows the receiver demodulator, to detect and correct, to a finite extent, errors as a result of the transmission process. Thus, improvement in BER performance is at the expense of an increase in transmission bit rate. The simplest error detection-only method used with digital binary messages is the parity check scheme. In the even-parity version of this scheme, the message to be transmitted is bundled into blocks of equal bits, and an extra bit is added to each block so that the total number of 1 s in each block is even. Thus, whenever the number of 1 s in a received block is odd, the receiver knows that a transmission error has occurred. Note, however, that this scheme can detect only an odd number of errors in each block. For error detection and correction, the addition of several

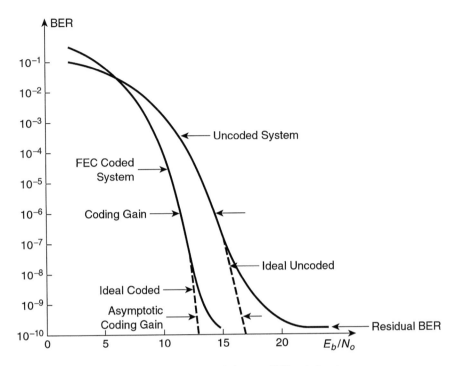

Fig. 6.1 Typical error performance of an uncoded versus FEC coded system

redundant (check) bits is required. The number of redundant bits is a function of the number of bits in errors that are required to be corrected.

FEC codes can be classified into two main categories, namely, *block codes* and *convolution codes*. However, though convolution codes are used in 3GPP 4G networks, they are not used in 5G NR, where all coding is block coding. We will thus only study block codes. We will first look at block coding in general and then turn our attention to those methods used specifically in 5G NR, namely, *low-density parity check* (LDPC) codes, *polar* codes, *Reed Muller* codes, *simplex* codes, and *repetition* codes.

6.3.2 Block Codes

In *systematic binary linear* block encoding, the input bit stream to be encoded is segregated into sequential message blocks, each k bits in length. The encoder adds r check bits to each message block, creating a codeword of length n bits, where $n = k + r$. The codeword created is called an (n, k) block codeword, having a block length of n and a coding rate of k/n. Such a code can transmit 2^k distinct codewords, where, for each codeword, there is a specific mapping between the k message bits and the r check bits. The code is *systematic* because, for all codewords, a part of the sequence in the codeword (usually the first part) coincides with the k message bits. As a result, it is possible to make a clear distinction in the codeword between the message bits and the check bits. The code is *binary* because its codewords are constructed from bits and *linear* because each codeword can be created by a linear *modulo 2 addition* of two or more other codewords. The following simple example [2] will help explain the basic principles involved in linear binary block codes.

Example 6.2 The Basic Features and Functioning of a Simple Linear Binary Code   Consider a (5, 2) block code where 3 check bits are added to a 2-bit message. There are thus four possible messages and hence four possible 5-bit encoded codewords. Table 6.1 shows the specific choice of check bits associated with the message bits. A quick check will confirm that this code is linear. For example, codeword 1 can be created by the modulo 2 addition of codewords 2, 3, and 4. How does the decoder work? Suppose codeword 3 (10011) is transmitted, but an error occurs in the second bit so that the word 11011 is received. The decoder will recognize that the received word is not one of the four permitted codewords and thus contains an error. This being so, it compares this word with each of the permitted codewords in turn. It differs in four places from codeword 1, three places from codeword 2, one place from codeword 3, and two places from codeword 4. The decoder therefore concludes that it is codeword 3 that was transmitted, as the word received differs from it by the least number of bits. Thus, the decoder can detect and correct an error.

The number of places in which two words differ is referred to as the *Hamming distance*. Thus, the logic of the decoder in Example 6.1 is, for each received word, select the codeword closest to it in Hamming distance. The minimum Hamming distance between any pair of codewords, d_{min}, is referred to as the *minimum distance* of the code. It provides a measure of the code's minimum error-correcting capability and thus is an indication of the code's strength. In general, the error-correcting capability, t, of a code is defined as the maximum number of guaranteed correctable errors per codeword and is given by:

$$t = \frac{d_{min} - 1}{2} \qquad\qquad (6.1)$$

where $\lfloor i \rfloor$ means the largest integer not to exceed i.

Table 6.1 A (5,2) block code

Codeword #	Codeword	
	Message bits	Check bits
1	0	000
2	01	110
3	10	011
4	11	101

An important subcategory of block codes is *cyclic block codes*. A code is defined as cyclic if any cyclic shift of any codeword is also a codeword. Thus, for example, if 101101 is a codeword, then 110110 is also a codeword, since it results from shifting the last bit to the first bit position and all other bits to the right by one position. This subcategory of codes lends itself to simple encoding algorithms. Further, because of their inherent algebraic format, decoding is also accomplished with a simple structure.

As indicated above, in a linear block code, each codeword can be created by a linear modulo 2 addition of two or more other codewords. Thus, it is possible to have a subset of all the codewords from which, by various combinations, all the other codewords can be formed. Such a subset is called a *basis*, and a so-called generator matrix, G, may be formed by writing the basis codewords bit by bit as the rows of a matrix. A linear block code of $m = 2^k$ information sequences of length k and codeword length of n have a $k \times n$ generator matrix G. Given G, each codeword c_i may be obtained by multiplying G by a different information sequence d_i, i.e.:

$$c_i = d_i G, 1 \leq I \leq m \tag{6.2}$$

The matrix G_x below is an example of a 2×5 generator matrix:

$$G_x = \begin{pmatrix} 0 & 0 & 1 & 1 & 1 \\ 1 & 1 & 0 & 0 & 1 \end{pmatrix}$$

For the information sequence 1 1, it can be shown that G_x generates, by modulo 2 matrix multiplication, the codeword 1 1 1 1 0. Based on this result, it is tempting to conclude that this is a systematic code. However, for the information sequence 0 1, the generated codeword is 1 1 0 0 1. The code is thus non-systematic.

Block decoding can be accomplished with *hard decision decoding*, where the demodulator outputs either ones or zeros as in Example 6.2. Here the codeword chosen is the one with the least Hamming distance from the received sequence. However, decoding can be improved by employing *soft decision decoding*. With such decoding the demodulator output is normally still digitized, but to greater than two levels, typically eight or more. Thus, the output is still "hard" but more closely related to the analog version and thus contains more information about the original sequence. Such decoding can be accomplished in a number of ways. One such way is to choose as the transmitted codeword the one with the least *Euclidian distance* from the received sequence. The Euclidian distance between sequences is, in effect, the root mean square error between them. To demonstrate the advantage of soft decision-based decoding over Hamming distance based decoding, a simple example of decoding via both methods is presented below.

Example 6.3 Demonstration of the advantage of soft decision decoding via Euclidian distance over hard decision decoding via Hamming distance

Consider an encoder that produces the four codewords in Table 6.1. This is a block encoder, and it serves the purpose of conveying in a straightforward fashion the basic concept and advantage of decoding using Euclidean distance versus Hamming distance as the decoding metric. Assume that codeword 2 (01110) is sent over a noisy channel in the form of the signal shown in Fig. 6.2a and, as a result, the demodulator analog output signal is as shown in Fig. 6.2b. This analog output leads to a hard decision output of 10010 and, with eight-level quantization, a soft decision output as indicated on the figure.

Let's first assume that decoding is based on hard decisions. If $d_H(r, n)$ represents the Hamming distance between the hard decision outputs of the received signal and codeword n, then simple comparison yields $d_H(r,1) = 2$, $d_H(r,2) = 3$, $d_H(r,3) = 1$, and $d_H(r,4) = 4$. Since $d_H(r,3)$ is the smallest Hamming distance, the decoder declares that the received codeword is codeword 3, i.e., 10011. It thus decodes in error.

Let's now assume that the decoder is using soft decisions and that $d_E(r,n)$ represents the Euclidean distance between the soft decision outputs of the received signal and codeword n. We compute the squared Euclidean distance, $d_E^2(r,1)$, by determining the error between the soft decision output and codeword 1 for each of the five

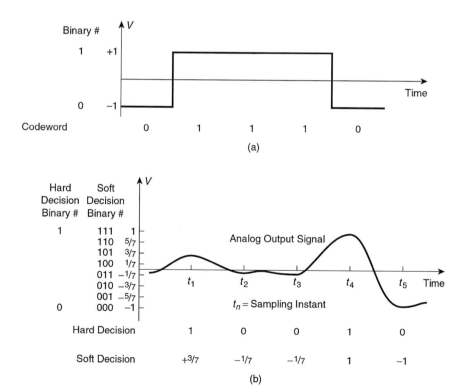

Fig. 6.2 Examples of hard and soft decoding. (**a**) Signal sent over noisy channel representing codeword 01110. (**b**) Demodulated analog output signal resulting from reception of signal in (**a**) plus noise and resulting hard and soft decisions

bits sent, squaring these errors, and then adding the squared values together. Since codeword 1 is 00000, its true output per bit would be $-1, -1, -1, -1, -1$, and thus the errors between its potential bit outputs and the received signal soft outputs are, sequentially, 1 3/7, 6/7, 6/7, 2, and 0. Thus, $d_E^2(r,1)$ is given by:

$$d_E^2(r,1) = (13/7)^2 + (6/7)^2 + (6/7)^2 + (2)^2 + (0)^2 = 7.51$$

Applying this same process to the other three codewords, we get:

$$d_E^2(r,2) = (13/7)^2 + (11/7)^2 + (11/7)^2 + (0)^2 + (0)^2 = 4.65$$
$$d_E^2(r,3) = (4/7)^2 + (6/7)^2 + (6/7)^2 + (0)^2 + (2)^2 = 5.80$$
$$d_E^2(r,4) = (4/7)^2 + (11/7)^2 + (11/7)^2 + (2)^2 + (2)^2 = 10.94$$

Since $d_E^2(r,2)$ is the smallest squared Euclidean distance, then $d_E(r,2)$ is the smallest Euclidean distance, and hence the encoder chooses codeword 2, thus making the correct decision. In effect, the decoder's decision is based on the fact that it can't, with much confidence, decide what are the first three bits that have been sent, but it can with a high confidence decide that the last two bits sent are 10. Since only codeword 2 had these last two bits, it decides, correctly, that codeword 2 was sent. Repetition (Sect. 6.3.6.1) and simplex (Sect. 6.3.6.2) codes used in 5G NR apply Euclidian distance in decoding.

Another form of soft decision decoding is via *Logarithmic Likelihood Ratios* (LLRs) where each received bit is processed as the probability, in a logarithmic form, of it being a 1. As we shall see in sections 6.3.4.2 and 6.3.5.3, this approach is used with 5G NR LDPC and polar codes.

6.3.3 Classical Parity Check Block Codes

Before we describe the features of LDPC codes, we review some of the features of classical parity check block codes. In such a code, each codeword is of a given length, n say, contains a given number of information bits, k say, and a given number of *parity check bits*, r say, and thus $r = n - k$. The structure can be represented by a *parity check matrix* (PCM), where there are n columns representing the digits in the codeword, and r rows representing the equations that define the code. Consider one such code, where the length n is 6, the number of information bits k 3, and hence the number of parity bits r is 3. The rate of this code is thus $k/n = 3/6 = 1/2$. We label the information bits V1 to V3 and the parity bits V4 to V6. The parity check equations for this code are shown in Eq. (6.3) below, where + represents modulo 2 addition:

$$V1 + V2 + V4 = 0$$
$$V2 + V3 + V5 = 0 \quad\quad\quad (6.3)$$
$$V1 + V2 + V3 + V6 = 0$$

The above equations can be represented in matrix form, as shown below in Fig. 6.3a, where each equation maps to a row of the matrix. This matrix is referred to as the PCM associated with Eq. (6.3).

Equation 6.3 can also be represented in graphical form. When done, such a graph is referred to as a *Tanner graph*. A Tanner graph is a bipartite graph, i.e., a graph which contains nodes of two different types and lines (also referred to as edges) which connect nodes of different types. The bits in the codeword form one set of nodes, referred to as *variable nodes* (VNs), and the parity check equations that the bits must satisfy form the other set of nodes, referred to as the *check nodes* (CNs). The Tanner graph corresponding to the PCM matrix above is shown in Fig. 6.3b.

Errors can be detected within limits in any received codeword by simply checking if it satisfies all associated parity check equations. However, block codes can only detect a set of errors if errors don't change one codeword into another. Further, even if this is not the case, they can only detect bit errors if the number of these errors is less than the minimum distance, d_{min}.

To not only detect bit errors but to also correct them, the decoder must determine which codeword was most likely sent. One way to do this is to choose the codeword closest in minimum distance to the received codeword. This method of decoding is called *maximum likelihood* (ML) *decoding*. For codes with a short number of information bits, this approach is feasible as the computation required is somewhat limited. However, for codes with thousands of information bits in a codeword such as can be found in 5G NR, the computation required becomes too excessive and expensive. For such codes, alternative decoding methods have been devised and will be discussed below.

6.3.4   Low-Density Parity Check (LDPC) Codes

Low-density parity check (LDPC) codes are employed in 5G NR. They are linear FEC codes and were first proposed by Gallager [3] in his 1962 Ph. D. thesis. Convolution turbo codes are used by 4G as its primary FEC codes. LDPC codes can provide, however, higher coding gains and lower error floors than turbo codes, and

Fig. 6.3 A PCM and associated Tanner graph. (**a**) PCM. (**b**) Tanner graph

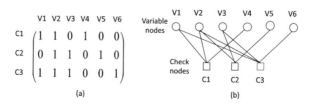

(a)   (b)

in the decoding process be computationally more efficient. LDPC codes are distinguished from other parity check codes by having parity check matrices where the percentage of 1 s is low, i.e., of low density, hence the nomenclature. This sparseness of 1 s results not only in a decoding complexity which increases only linearly with code length but also a minimum distance which increases only linearly with code length.

LDPC codes are said to be either regular or irregular. A LDPC code is regular if all the VNs have the same degree, i.e., they are connected to the same number of CNs, and all the CNs have the same degree, i.e., they are connected to the same number of VNs. When this is the case, then every code bit is contained in the same number of equations, and each equation contains the same number of code bits. Irregular codes relax these conditions, allowing VNs and CNs of different degrees. Irregular codes, it has been found, can provide better performance than regular ones.

Figure 6.4a shows the PCM of a simple LDPC where $n = 12$, and Fig. 6.4b shows the associated Tanner graph. It will be noted that this is a regular LDPC code, where from the PCM perspective each code bit is contained in 3 equations and each equation involves 4 code bits and from the Tanner graph perspective, each bit node has 3 lines connecting it to parity nodes, and each parity node has 4 lines connecting it to bit nodes. We note that in the PCM, there are 108 positions in all of which only 36%, or 33%, are ones.

6.3.4.1 Encoding of Quasi-cyclic LDPC Codes

5G NR uses a class of LDPC codes called *Quasi-cyclic* (QC) LDPC codes [4]. These codes are used in support of channels that transmit user data in both downlink and uplink directions. They are also used in downlink paging channels. With these codes, encoding and decoding hardware implementation tends to be easier than with other type of LDPC codes, achieving this without measurably degrading the relative performance of the code. The PCM of a QC LDPC codes is defined by a small graph, called a *base graph* or *protograph*. The base graph, U say, is transformed into the PCM, H say, by replacing each entry in U with a cyclically shifted to the right version of a Z × Z identity matrix, I say. Here, Z is referred to as the *lifting factor*,

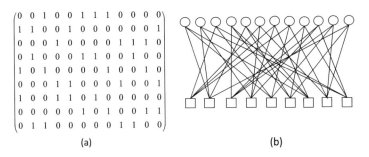

(a)   (b)

Fig. 6.4 PCM and Tanner graph for an LDPC code where $n = 12$. (**a**) PCM. (**b**) Tanner graph

as the larger it is, the larger the size of H. Base graph entries are not binary but rather range from -1 to $(Z-1)$. By convention, -1 means a matrix with all 0 entries. The entries 0 to $(Z-1)$ represent the possible cyclically shifted versions of I. Example 6.4 below will demonstrate the construction of a QC LDCP code.

Example 6.4 Construction of a QC LDPC Code Consider a QC LDPC code where the identity matrix I is as below:

$$I = \begin{pmatrix} 1 & 0 & 0 \\ 0 & 1 & 0 \\ 0 & 0 & 1 \end{pmatrix}$$

Then the base graph entries are given by:

$$-1 = \begin{pmatrix} 0 & 0 & 0 \\ 0 & 0 & 0 \\ 0 & 0 & 0 \end{pmatrix} \; 0 = \begin{pmatrix} 1 & 0 & 0 \\ 0 & 1 & 0 \\ 0 & 0 & 1 \end{pmatrix} \; 1 = \begin{pmatrix} 0 & 1 & 0 \\ 0 & 0 & 1 \\ 1 & 0 & 0 \end{pmatrix} \; 2 = \begin{pmatrix} 0 & 0 & 1 \\ 1 & 0 & 0 \\ 0 & 1 & 0 \end{pmatrix} \; 3 = \begin{pmatrix} 1 & 0 & 0 \\ 0 & 1 & 0 \\ 0 & 0 & 1 \end{pmatrix}$$

Thus, if we have a base graph U given by:

$$U = \begin{pmatrix} 3 & -1 & -1 & 1 \\ -1 & 0 & -1 & 2 \end{pmatrix}, \quad \text{then } H = \left(\begin{array}{ccc|ccc|ccc|ccc} 1 & 0 & 0 & 0 & 0 & 0 & 0 & 0 & 0 & 0 & 1 & 0 \\ 0 & 1 & 0 & 0 & 0 & 0 & 0 & 0 & 0 & 0 & 0 & 1 \\ 0 & 0 & 1 & 0 & 0 & 0 & 0 & 0 & 0 & 1 & 0 & 0 \\ \hline 0 & 0 & 0 & 1 & 0 & 0 & 0 & 0 & 0 & 0 & 0 & 1 \\ 0 & 0 & 0 & 0 & 1 & 0 & 0 & 0 & 0 & 1 & 0 & 0 \\ 0 & 0 & 0 & 0 & 0 & 1 & 0 & 0 & 0 & 0 & 1 & 0 \end{array} \right)$$

We note that the key to high performance of a QC LDPC code is the construction of the base graph. Though the number of identity matrix cyclic permutations is Z, in practice the number of permutations used is restricted to simplify implementation.

5G NR LDPC codes specify two base graphs in order to cover a large range of information payloads and rates. Figure 6.5 shows the general structure of both these base graphs. As can be seen in the figure, the base graph consists of five submatrices labeled A, D, O, E, and I [5]:

- A corresponds to the systematic bits.
- D is a square matrix and corresponds to the first set of parity bits.
- O is a zero matrix.
- E corresponds to single parity check rows.
- I is an identity matrix.

Fig. 6.5 General structure
of 5G NR base graphs

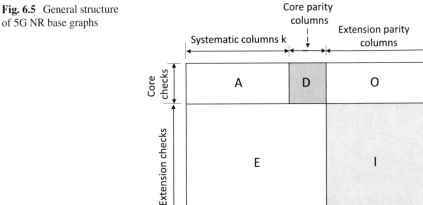

The combination A and D is referred to as the kernel. Parts O, E, and I are referred to as extensions.

Base graph 1 (BG1) is specified to accommodate large payloads and high rates. It is of dimension 46 × 68 and contains 22 systematic information columns. With a maximum lift factor of 384, this results in a maximum information payload of 22 × 384 = 8448 bits. Nominal rate is 8/9, but, by adding additional parity bits to the base graph, the rate can be decreased to as low as 1/3 without puncturing or repetition. This low rate aids in permitting extended coverage and the meeting of the cell-edge throughput goal of 100 MB/s.

Base graph 2 (BG2) is specified to accommodate smaller payloads and lower rates. It is of dimension 42 × 52 and contains either 6, 8, 9, or 10 systematic information columns depending on the payload size. With the same maximum lift factor as base graph 1 of 384, this results in a maximum information payload of 10 × 384 = 3840 bits. Nominal rates vary between 2/3 and 1/5 without puncturing or repetition. The choice between base graph 1 and base graph 2 is a function of transport block size and the code rate chosen for the initial transmission and shown graphically in Fig. 6.6.

6.3.4.2   Decoding of LDPC Codes

A big distinguishing feature between LDPC codes and classical block codes is how they are decoded. Unlike classical codes that are usually of short length and decoded via ML decoding, LDPC are decoded iteratively.

LDPC codes are decoded using *message-passing* algorithms [6] since their functioning can be described as the passing of messages along the lines of the Tanner graph. Each node on the Tanner graph works in isolation, having access only to the information conveyed by the lines connected to it. The message-passing algorithms

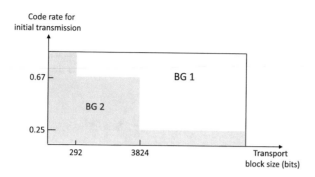

Fig. 6.6 Choice of LDPC base graph

create a process where the messages pass back and forth between the bit nodes and check nodes iteratively. For optimum decoding the messages passed are estimates of the probability that the codeword bit information passed is 1. Each estimate is in the form of a *Logarithmic Likelihood Ratio* (LLR), where, for a codeword bit b_i, $\text{LLR}(b_i) = \ln \text{prob.}(b_i = 0) - \ln \text{prob.}(b_i = 1)$. A positive LLR indicates a greater confidence that the associated bit is of value 0, while a negative LLR indicates a greater confidence that the bit value is 1. The magnitude of the LLR expresses the degree of confidence. Decoding as described above is termed *belief propagation* decoding and proceeds as follows:

1. Each codeword is outputted from the channel not as hard outputs (1s or 0s) but rather as soft outputs. These soft outputs are converted into initial estimates in the form of LLRs.
2. Each bit node sends its initial estimate to the check nodes on the lines connected to it.
3. Each check node makes new estimates of the bits involved in the parity equation associated with that node and sends these new estimates via the connecting lines back to the associated bit nodes.
4. New estimates at the bit nodes are sent to the check nodes and process steps 3 and 4 repeated until a permitted codeword is found or the maximum number of permitted iterations reached.

6.3.5 Polar Codes

Polar codes are block codes and were invented by Arikan and disclosed in 2009 [7]. They are the first error-correcting codes that are theoretically able to achieve the capacity of a *binary discrete memoryless channel* (Bi-DMS). By memoryless channel we mean one where the output signal at a time t is only determined by the input signal at time t and consequently not dependent on the signal transmitted before or after t. By capacity, we mean capacity as defined by Shannon [8], who showed that it was theoretically possible for a communication system to transmit information with an arbitrarily small probability of error if the information rate R is less than or equal to a rate C, the *channel capacity*, where, for a channel in which the noise N, is bandlimited Gaussian, B is the channel bandwidth, and S is the signal power, then C is given by the Shannon Hartley theorem [8]:

$$C = B \log_2 \left(1 + \frac{S}{N}\right) \text{bits} / \text{s} \tag{6.4}$$

By permitting the achievement of channel capacity, polar codes, in theory, permit the transmission of bits at rate C, the highest rate possible with negligible error. In practice, however, C is not attainable as it requires unreasonably large block lengths. This disadvantage is offset, however, by the fact that encoding and decoding operations can be performed with low complexity in a deterministic recursive fashion.

6.3.5.1 Channel Polarization

Polar codes are able to approach channel capacity by employing *channel polarization*, hence its nomenclature. With channel polarization, channels are constructed to be mostly of high capacity and low capacity. As more and more such channels are constructed recursively, the high capacity channels get more so, and the low capacity ones get more so. The number of polarized channels used is a function of the code length. Thus, as code length increases, the number of polarized channels and hence polarization increases. A key to polar coding, as we will see below, is that the high capacity channels are used to transmit the information bits, and the low capacity ones used to transmit "frozen" bits, normally set to zero.

To aid in describing channel polarization, we first introduce, for those unfamiliar with it, the *binary erasure channel* (BEC). The BEC can transmit at any one time only one of two symbols, a 0 or a 1. A model of the BEC is shown in Fig. 6.7. When a bit is inputted to the channel, the channel outputs either the sent bit or a message that it was not received, i.e., erased (erasure symbol given by?). If the probability of erasure is p_e say, then the probability of the bit being outputted is $1 - p_e$. Also, it can be shown that the capacity of a BEC is $1 - p_e$.

To understand the polarization effect, we examine Fig. 6.8a, which shows a two input and two output channel combiner, employing two BECs, W_1 and W_1, each with capacity $C(W_1)$. These two BECs are combined with the aid of a modulo 2

Fig. 6.7 Binary
erasure channel

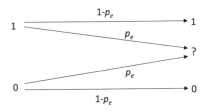

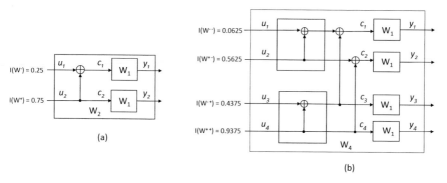

Fig. 6.8 (**a**) Two-channel polarizing combiner (**b**) Four-channel polarizing combiner

adder to form the compound channel W_2 with a total capacity of $2C(W_1)$. It can be shown that W_2 can be treated as being *split* into two channels W^+ and W^-, with U_1 being the input to W^- and U_2 being the input to W^+ [9]. With this split structure, it can be further shown that the capacity of W^+ is equal to $2C(W_1) - C(W_1)^2$, and the capacity of W^- equal to $C(W_1)^2$ [9]. Thus, though the capacity of each of these new channels is different, total capacity of the system is preserved. If each W_1 was to have an erasure probability of 0.5, then they would each have a capacity $C(W_1)$ of 0.5, and the capacity of W^+ would be 0.75 and that of W^- would be 0.25. We thus see that under this scenario, the channels have started to polarize. The key to polar coding is that as the number of channels increases, the degree of polarization increases. Figure 6.8b shows a four-channel combiner labeled W_4, created by combining two W_2 compound channels and having the individual channel capacities as shown. As this combining process is repeated, the capacities of more and more channels migrate towards either one or zero. Importantly, the polarizing effect works not only for a set of BECs, but also for AWGN channels, where polarization not only addresses capacity but also BER. However, determining the reliability order of AWGN channels, and hence what channels to assign to information bits, is more complex than for BECs.

An important feature of polar codes compared to other FEC codes is that they have been shown analytically to not suffer from an error floor.

6.3.5.2   Encoding of Polar Codes

To create a polar encoder, we must know the code block length n to be transmitted, where n must be a power of two, and the number of information bits k per code block. The n-k non-information bits are referred to, as mentioned above, as frozen bits and are normally set to 0. The encoder consists of the compounded polarizing encoder W_N. Given the calculated capacities of the individual channels, the k bits are assigned to the channels with the highest capacities and hence the lowest probabilities of error. Figure 6.9 shows a polar encoder for $n = 8$, rate $R = 1/2$, and hence

$k = 4$ and based on transmission over BEC channels of probability of erasure of ½. Observe that the information bits are assigned to the four channels with the highest capacities.

6.3.5.3   Decoding of Polar Codes

The method normally used to decode polar codes is called *successive cancellation* (SC) decoding [9], a method that is effective enough to achieve capacity at infinite code length. Here the decoder makes hard bit decisions one at a time, using as the inputs to its computation both the soft information received from the channel in the form of LLRs, as well as the hard decisions made on the previously decoded bits. The algorithm used to determine the value of the bit being decoded is quite complex and involves many LLR computations. If a bit is frozen, then it sets its value to 0. To see how this works, let's consider the decoding of the codes produced by the encoder shown in Fig. 6.9. The decoding proceeds as follows:

Stage 1. Decode U_1. Frozen bit. Hence $U_1 = 0$.
Stage 2. Decode U_2. Frozen bit. Hence $U_2 = 0$.
Stage 3. Decode U_3. Frozen bit. Hence $U_3 = 0$.
Stage 4. Decode U_4. Information bit: Use Y_1 through Y_8, and U_1 through U_3 to decode.
Stage 5. Decode U_5. Frozen bit. Hence $U_5 = 0$.
Stage 6. Decode U_6. Information bit: Use Y_1 through Y_8, and U_1 through U_5 to decode.
Stage 7. Decode U_7. Information bit: Use Y_1 through Y_8, and U_1 through U_6 to decode.
Stage 8. Decode U_8. Information bit: Use Y_1 through Y_8, and U_1 through U_7 to decode.

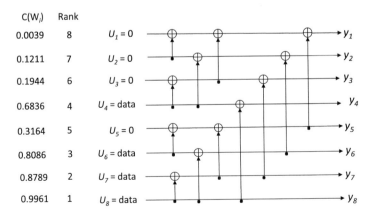

Fig. 6.9 Eight-channel polar encoder

An example of a tree representation of SC decoding for a codeword produced by the encoder of Fig. 6.9 is shown in Fig. 6.10. For simplicity purposes, imaginary-computed likelihood values (LLR values in real systems) of individual bits being either 1 or 0 are shown beside each node, and the associated bit decision is shown next to the preceding tree branch. The path computed by the algorithm is shown in solid lines and rejected paths shown in broken lines. The stages shown are for the stages outlined above. For stages 1, 2, and 3, the decoded decisions are all 0 as these are frozen bits. At stage 4 the likelihood of 1 is higher than that of 0, so 1 is chosen, and the path to the left is terminated. Stage 5 is the decoding of a frozen bit; hence the decision is zero. Stages 6, 7, and 8 are decoded as shown, leading to and output sequence of 0 0 0 1 0 0 0 1 and hence an information sequence of 1 0 0 1(bits number 4, 6, 7, and 8). In general, the decision made on any bit is influenced by all previous bit decisions. If there is an incorrect bit decision, it cannot be corrected later and thus can result in a cascade of errors in subsequent bits.

Though easy to implement, a concern with SC decoding is its relatively high latency, resulting primarily from the fact the information bits are decoded one by one. Further, the performance of practical finite length polar codes with SC decoding is noticeably worse than other competitive FEC codes [10]. This latter problem can be addressed with *successive cancellation list* (SCL) *decoding* [9] which substantially improves the BLER performance of SC.

SCL strives to overcome the limitation of premature decisions taken by SC decoding by employing a list of possible bit sequences, of length L, as it moves from one decoding stage to the next. The list is temporarily doubled at the beginning of each decoding stage, the likelihood of all 2 L paths compared, and only the L paths with the highest likelihoods retained and considered at the next decoding stage. At the end of the process, the SCL outputs the sequence with the highest likelihood. By utilizing a list, the decoder forestalls an early decision which may be incorrect.

An example of a tree representation of SCL decoding for a codeword produced by the encoder of Fig. 6.9 is shown in Fig. 6.11 for a list size of L = 2. As was done for the SC decoder, imaginary-computed likelihood values of individual bits being

Fig. 6.10 SC decoding of a polar code encoded by the encoder of Fig. 6.8

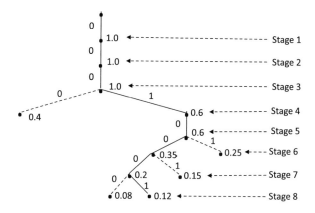

either 1 or 0 are shown beside each node, and the associated bit decision shown next to the preceding tree branch. The path computed by the algorithm is shown in solid lines and rejected paths shown in broken lines. For stages 1, 2, and 3, the decoded decisions are all 0 as these are frozen bits. At stage 4, though the likelihood of 1 is higher than that of 0, the decoder does not make a decision but keeps both paths under consideration. Stage 5 is the decoding of a frozen bit hence the decision is zero. At stage 6, the encoder computes the likelihood of all four possible paths going forward. However, as the list size is 2, it chooses the two paths with the highest likelihoods and discards the other two. This process of pruning is repeated at stage 7, and at stage 8 it ends up with four possible paths. Here it chooses the path of highest likelihood, the one ending with a likelihood of 0.15. This leads to an output sequence of 0 0 0 0 0 1 0 0 and hence an information sequence of 0 1 0 0. It will be observed that, with the additional likelihoods given relative to SC decoding tree of Fig. 6.10, the SCL decoded information sequence is quite different to the SC decoded one.

To further improve BLER performance, polar codes can employ *CRC-aided SCL* (CA-SCL) [10]. It was found that when errors occurred with SCL decoding, the correct sequences were usually in the final L-sized list but that it was not the codeword with the highest likelihood and hence not selected in the last step of the decoder. With a CRC added to the codeword, however, the codeword in the list that passes the CRC test can be declared the correct codeword. CA-SCL improves the BLER performance of SCL polar codes but at the expense of a slight decrease in the code rate. A main concern of SC decoding is large latency. It has been found that this problem can be alleviated by distributing the CRC information bits within the information bits rather than at the end of the information bits.

6.3.5.4 5G NR Polar Codes

Polar codes are used in 5G NR in support of downlink control channels, uplink control channels where the information block lengths are greater than 11 bits, and broadcast channels. In control channels, a smaller amount of information bits with

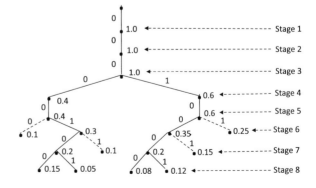

Fig. 6.11 SCL decoding of a polar code encoded by the encoder of Fig. 6.8

a shorter codeword length is typically transmitted compared to user data channels. Further, low code rates are typically used, and good performance in a low block error rate environment is required. 5G polar codes are designed to support these control channel requirements.

For the downlink, 5G polar codes are of code lengths 2^n for $7 \leq n \leq 9$. Thus maximum length is $2^9 = 512$, and the minimum length is $2^7 = 128$. For the uplink, code lengths are for $5 \leq n \leq 10$, and thus maximum length is $2^{10} = 1024$, and minimum length is $2^5 = 32$. 5G NR polar codes employ CRC-aided SCL decoding.

6.3.6 Repetition, Simplex, and Reed Muller Block Codes

For uplink control information (UCI), 5G NR uses binary repetition, simplex, or Reed Muller codes depending on the information code block length.

6.3.6.1 Repetition Codes

If the UCI block length is 1, 5G NR uses repetition codes for UCI conveyance (Sect. 5.3.3.1 of [11]). Repetition codes are one of the simplest types of error-correcting codes. Here, the binary message to be transmitted is simply repeated several times with the hope that if corrupted during transmission, only a minority of bits will be corrupted. Decoding would then be done by a majority decision. A binary repetition code of length n is a $(n, 1)$ code with two codewords, one being all zeros, the other all ones. Its minimum Hamming distance d_{min} is n, and it can thus correct up to $(n - 1)/2$ errors in any codeword. As an example of encoding, consider the case where the information bits 101 are to be transmitted with a code length of 3. Then the encoded message would be 111 000 111.

In 5G NR, the amount of repetition is a function of the modulation order. If fact, if the modulation is $\pi/2$ BPSK, there is no repetition, whereas, if the modulation is 256-QAM, the information bit is repeated seven times to create a codeword 8 bits long. Unlike standard repetition coding, in 5G NR the repeated information bits are scrambled "in a way that maximizes the Euclidean distance of the modulation symbols carrying the information bits."

6.3.6.2 Simplex Codes

If the UCI block length k is 2, 5G NR uses simplex codes for UCI conveyance (Sect. 5.3.3.2 of [11]). Such codes are error-detecting or error-correcting linear cyclic block codes. A detailed description of these codes will not be covered here, but we note that for a binary simplex (n,k) code:

$$n = 2^k - 1$$
$$d_{min} = 2^{k-1}$$

Thus, for the case where the number of information bits k equals two, we have $n = 3$ and $d_{min} = 2$.

For basic 5G NR simplex encoding, if the two information bits are c_0 and c_1, then the third encoded bit c_2 is defined as $c_2 = (c_0 + c_1)$, where the addition is modulo 2. Thus, the four possible encoded words are 0 0 0, 0 1 1, 1 0 1, and 1 1 0. When the modulation is $\pi/2$ BPSK, there are three encoded bits, $[c_0\ c_1\ c_2]$. When the modulation is QPSK, the three basic encoded bits are repeated to give six encoded bits, $[c_0\ c_1\ c_2\ c_0\ c_1\ c_2]$. When the modulation is 16, 64, or 256 QAM, there are 12, 18, and 24 encoded bits, respectively. Here, however, the repetition of c_0, c_1, and c_2 includes scrambled bits to, as with repetition encoding, maximize the Euclidian distance of the modulation symbols carrying the information bits.

6.3.6.3   Reed Muller Codes

If the UCI block length k is bounded by $3 \leq k \leq 11$, then 5G NR uses *Reed Muller* (RM) codes [2] for UCI conveyance (Sect. 5.3.3.3 of [11]). RM codes are a class of linear cyclic block codes that can be encoded and decoded with simple algorithms. For all (n,k) RM codes, n is always a power of 2. Thus, if this power is m, then $n = 2^m$.

Generator matrices for RM codes come in different orders, r. The generator matrix of the zero-order ($r = 0$) RM code, G_0, consists of one row containing all ones, making this simply a repetition code. The generator matrix of the first order ($r = 1$) code is created by adding below G_0 a new matrix, G_1, which contains m rows, whose columns consist of all n possible distinct binary sequences put in natural binary order. The generator matrix of the second-order ($r = 2$) code is created by adding below G_1 a new matrix, G_2, which contains $m(m - 1)/2$ rows consisting of all possible logical AND combinations of two of the additional first order rows. (Note that logical AND, denoted by "&," is the same as modulo 2 multiplication). In general, the rth order generator adds to all combinations of r of the first order rows to the $(r - 1)$th order generator. Figure 6.12 below shows the second-order generator matrix for a length 8 ($m = 3$) RM code.

The minimum distance d_{min} of a rth order RM code is $2^{m - r}$. RM codes can be decoded using majority logic decoding [12], a method that is not very efficient but straightforward to implement.

In 5G NR Reed Muller codes used are $(32, k)$ codes. For a 32 length RM code, $m = 5$, the first order code is a $(32, 6)$ code, and the second-order code is a $(32,16)$ one. This suggests that of the 3 to 11 block lengths covered by RM coding, only a length of 6 can be accommodated. This limitation is addressed by utilizing a *subcode* [13] of the second-order RM code. This subcode has an 11×32 generator matrix (Table 5.3.3.3–1 of [11]), the design of which is beyond the scope of this text. Suffice it to say that the first column contains all ones, the following 5 columns

Fig. 6.12 Second-order   Row #
generator for length
8 RM code

$$
\begin{array}{c}
G_0 \\
\\
G_1 \\
\\
\\
G_2 \\
\\
\end{array}
\left(
\begin{array}{cccccccc}
1 & 1 & 1 & 1 & 1 & 1 & 1 & 1 \\
\hline
0 & 0 & 0 & 0 & 1 & 1 & 1 & 1 \\
0 & 0 & 1 & 1 & 0 & 0 & 1 & 1 \\
0 & 1 & 0 & 1 & 0 & 1 & 0 & 1 \\
\hline
0 & 0 & 0 & 0 & 0 & 0 & 1 & 1 \\
0 & 0 & 0 & 0 & 0 & 1 & 0 & 1 \\
0 & 0 & 0 & 0 & 0 & 0 & 1 & 1 \\
\end{array}
\right)
\begin{array}{c}
0 \\
1 \\
2 \\
3 \\
1\&2 \\
1\&3 \\
2\&3 \\
\end{array}
$$

consist of all possible distinct 32 binary sequences, but not in natural binary order, and the last 5 columns consist of 20 distinct binary sequences and 6 pairs of distinct binary sequences. For $k = 11$, the full 11×32 subcode generator matrix is used in encoding. For $k < 11$ only the first k rows of the subcode matrix are employed.

6.4   Puncturing

Puncturing is the process of discarding some of the bits of an error correction codeword prior to transmission. By employing puncturing, and thus creating punctured codes, it is possible to increase the code rate of a given code. To understand how puncturing works, consider the case where we want to create a rate ¾ code from a rate ½ code. For the rate ½ code, for every three 3-bit input sequence, we have a 6-bit output. To create the rate ¾ code, we simply delete 2 of the 6 output bits, thus giving us 4 output bits for every 3 input ones. The performance of this punctured code is dependent on which bits were deleted. The rate ¾ punctured code can be decoded using same decoder as required for original unpunctured rate ½ code. To use the rate ½ decoder, the rate ¾ punctured code is transformed back into a rate ½ structure by simply inserting dummy symbols (1s or 0s) into positions where bits were deleted before decoding. The dummy bits result in an impairment of the rate ½ code correcting capability. However, the impaired capability is normally no less than that which would have been achieved had an unpunctured rate 3/4 code been employed in the first place. Punctured codes allow dynamic selection of code rate based on actual propagation conditions.

6.5   Block Interleaving

Block interleaving is a technique applied to mitigate the impact of error bursts. A simple example will illustrate how *block interleaving* addresses error bursts in a simple block code.

Example 6.4 How block interleaving impacts the decoding of signals corrupted with error bursts.

In this example, an encoder creates the original 4-bit codewords shown in Fig. 6.13a. These codewords are fed to an interleaver that creates the interleaved words shown in Fig. 6.13b. Assume these interleaved words are then transmitted over a wireless noisy channel. As a result, a burst of five contiguous errors appears on the demodulated interleaved words, as indicated in Fig. 6.13b. Note, however, that when de-interleaved as shown in Fig. 6.13c, the five errors are now spread over the five original codewords. Assuming a decoder that can correct just one bit per codeword, it can, nonetheless, decode the original five codewords without error. Without interleaving, a burst of five contiguous errors would have caused errors in two codewords, which would have been beyond the capability of the decoder to eliminate.

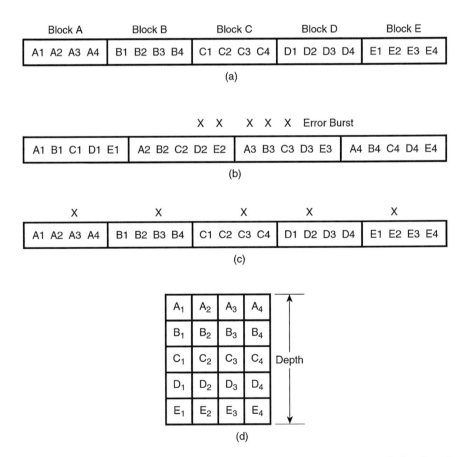

Fig. 6.13 Block interleaving. (**a**) Original codewords. (**b**) Interleaved words. (**c**) Deinterleaved words. (**d**) Two-dimensional array

The number of coded blocks (codewords) involved in the interleaving process is referred to as the *interleaving depth*. Thus, our simple interleaver in Example 6.4 has an interleaving depth of five. The larger the interleaving depth, the longer the burst of contiguous errors that can be corrected but the greater the delay introduced. A block interleaver consists of a structure that supports a two-dimensional array, of width equal to the codeword length and depth equal to the interleaving depth. For the interleaver of Example 6.4, the data is fed in row by row until the array is full, as shown in Fig. 6.13d, and then read out column by column, resulting in a permutation of the order of the data. At the receive end of data transmission, the original data sequence is restored by a corresponding de-interleaver. A common designation for an interleaver is π and that for its corresponding de-interleaver π^{-1}.

6.6 Automatic Repeat Request (ARQ)

The automatic repeat request (ARQ) error control method is used with packet data systems where data transmission is not a continuous stream. In this method, the receiver detects packet errors, typically via CRC, but makes no attempt to correct packets received in error. Rather, if a packet error is not detected, an ACK message sent to transmitter, which then sends the next packet. However, if a packet error is detected, a NACK message is sent to the transmitter which then retransmits the packet in error. ARQ error control is not very efficient, since if just one bit in a codeword is in error, a retransmission of the entire codeword is required.

6.7 Hybrid ARQ (HARQ)

Hybrid ARQ (HARQ) is an error control scheme that employs both FEC and ARQ. By combining advantages of both these schemes, much better performance than ARQ is achieved, especially in time-varying fading channels and channels experiencing fluctuating interference. There are two main types of HARQ schemes: type-I and type-II. In the type-I scheme, the receiver discards any packet received in error that it is unable to correct and requests a retransmission. This process continues until an error-free packet is available or until a maximum number of retransmission attempts have taken place. In the type-II scheme, which is the more efficient of the two, any packet that cannot be successfully decoded is saved for possible further use, and a retransmission is requested. There are two forms of type-II scheme: *chase combining* and *incremental redundancy*.

Chase combining (CC) is the simpler form of the type-II scheme. Here, when an error is detected at the receiver, a NACK is sent back to transmitter, but the packet in error stored in a buffer memory is not discarded. If the retransmitted packet, which is an identical copy of original, is again in error, then it is combined with previous packet in an attempt to correct the error. New packets are transmitted with the same scheme being applied until either the error is corrected, or the maximum number of retries specified is reached.

Incremental redundancy (IR), the more complex form of the type-II scheme, is similar to CC, but here each retransmitted packet contains a different mix of information bits and parity bits. Transmitted packets are identified by version. The first packet, which typically contains all the information bits in addition to some parity bits, is referred to as redundancy version 0 or RV0 and is followed by other redundancy versions.

The combining schemes used in CC and IR where, as we have seen, the receiver combines the received signals from multiple transmission attempts, are called "soft-combining" schemes, and bits combined are called "soft bits." HARQ with incremental redundancy is the primary way of handling retransmissions in NR.

6.8 Adaptive Modulation and Coding (AMC)

In a mobile cell, the further a mobile unit (MU) is from the base station (BS) the lower the average signal level received by both the MU and the BS. In addition, the further the MU from the BS, the larger the fading that is likely to occur on the BS/MU path. If a high-order modulation, for example, 64-QAM, is used on the transmission from one MU and a low-order from another, for example, QPSK, the MU with the high-order modulation will be able to send more data in a given bandwidth than the one with the low-order modulation. However, at the BS, the received power from the high-level MU must be greater than that from the low-level one for similar BER performance, as BER performance decreases as modulation level increases. This suggests that for similar BER performance on individual MU to BS links, modulation should be function of path length, assuming equal transmitter power on all MUs. High-level modulation could be employed on short paths, low-level on long ones, and thus the greater throughput the shorter the path. For scheme to be truly effective, however, it should be implemented in downstream direction as well.

In adaptive modulation, modulation is adjusted automatically per BS/MU link, independently on downstream and upstream directions, so as to optimize trade-off between capacity and reach. Modulation is typically adjusted on a data burst-by-data burst basis. In addition to optimizing throughput and coverage area, adaptive modulation also helps in combating co-channel interference. It does this by decreasing modulation complexity and hence increasing interference resistance whenever performance starts to degrade due to such interference. Instantaneous measurements are made of the S/N and C/I and modulation complexity varied dynamically. For example, if a MU has good SNR because its path is not fading, more complex modulation is used. If, on the other hand, the BS has poor C/I because its received signal is fading, and an interferer is present, less complex modulation is used.

An improvement on adaptive modulation is adaptive modulation and coding (AMC), where not only is modulation adjusted dynamically but coding as well. AMC is used in 5G NR. It allows the current modulation-coding scheme to be matched to current channel conditions for each user. With AMC, the power of the transmitted signal is held constant over a frame interval, and the modulation-coding format changed burst-by-burst within the frame to match users current received signal conditions. To effect AMC, each transmitter must have knowledge of the path condition; hence feedback from receiver to transmitter is required.

In a mobile system, a difference is inevitable between channel status when reported (feedbacked) and when applied. HARQ is very effective in addressing the difference problem. Thus, AMC combined with HARQ provides robust transmission over a time-varying channel.

6.9 Summary

In this chapter we focused on error detection and error correction coding as applied in 5G NR, including the very important low-density parity check (LDPC) codes, polar codes, and hybrid automatic repeat request (Hybrid ARQ). Such detection and correction are crucial to achieving bit probability of error performance required of 5G systems. Channel coding has been from interception an ever-improving technology. However, performance today is so close to the theoretical optimum predicted by Shannon that further improvement is likely to be limited. That said, measurable improvement in the areas of latency and easy of realization are certainly possible, so it is reasonable to expect the evolution to continue.

References

1. Peterson WW, Brown DT (1961) Cyclic codes for error detection. In: Proceedings of the IRE, vol 49, Jan 1961
2. Burr A (2001) Modulation and coding for wireless communications. Pearson Education, Harlow
3. Gallager RG (1963) Low density parity codes. MIT Press, Cambridge, MA
4. Fossorier MPC (2004) Quasi-cyclic low-density parity-check codes from circulant permutation matrices. IEEE Trans Inf Theory 50(8):1788–1793
5. Li H et al (2018) Algebra-Assisted Construction of Quasi-Cyclic LPDC Codes for 5G New Radio. IEEE Access, Special Section on Advances in Channel Coding for 5G and Beyond, Digital Object Identifier https://doi.org/10.1109/ACCESS.2018.2868963, Sept 2018
6. Mackay DJC (1999) Good error-correcting codes based on very sparse matrices. IEEE Trans Inf Theory 45:399–431
7. Arikan E (2009) Channel Polarization: a method for constructing capacity achieving codes for symmetric binary-input memoryless channels. IEEE Trans Inf Theory 55(7):3051–3073
8. Taub H, Schilling DL (1968) Principles of communication systems, 2nd edn. McGraw Hill, Inc., New York
9. Arikan E (2008) Channel polarization: a method for constructing capacity-achieving codes. In: Proceedings of IEEE international symposium on information theory, Toronto, July 2008, pp 1173–1177
10. Tal I, Vardy A (2015) List decoding of Polar codes. IEEE Trans Inf Theory 61(5):2213–2226
11. 3GPP Technical Specification TS 38.212 (2018) version 15.2.0, Release 15
12. Cooke B (1999) Reed-Muller error correcting codes. MIT Undergraduate Journal of Mathematics
13. Van Wonterghem J et al (2018) On Construction of Reed-Muller Subcodes. IEEE Commun Lett 22(2):220–223

Chapter 7
Multi-carrier-Based Multiple-Access Techniques

7.1 Introduction

Multiple access, in a mobile environment, implies two-way communication between a base station (BS) and multiple surrounding mobile units (MUs). Multi-carrier implies communication via a plurality of radio-frequency carriers, normally referred to as sub-carriers. In this chapter we will explore multi-carrier-based multiple-access techniques as applicable to 5G NR. First, we will review orthogonal frequency division multiplexing (OFDM). OFDM is a multi-carrier technique but not normally by itself a multiple-access one. On its own it supports point-to-point communications. Its basic structure does, however, lend itself to form the basis of multiple-access techniques. Two such techniques are employed by 5G NR, namely, orthogonal frequency division multiple access (OFDMA) and discrete Fourier transform spread OFDM (DFTS-OFDM). Having reviewed OFDM and some of its implementation issues, OFDMA and DFTS-OFDM are addressed. Next, three other multiple access schemes that were considered for application in 5G NR but ultimately not supported are introduced. Finally, we will take a high-level view of non-orthogonal multiple access (NOMA). NOMA, as envisaged for possible use in 6G or beyond, and under study by 3GPP, still applies OFDM principles, but in such a way as to have transmitted signals that are non-orthogonal to each other. The motivation to explore the use of NOMA is that it holds the promise of greater network capacity, i.e., greater data throughput capability.

The original version of this chapter was revised. The correction to this chapter is available at https://doi.org/10.1007/978-3-030-51441-9_12

© Springer Nature Switzerland AG 2020, Corrected Publication 2021
D. H. Morais, *Key 5G Physical Layer Technologies*,
https://doi.org/10.1007/978-3-030-51441-9_7

7.2   Orthogonal Frequency Division Multiplexing (OFDM)

7.2.1   OFDM Basics

Orthogonal frequency division multiplexing (OFDM) is not a modulation technique though it is often loosely referred to as such. Rather, it is a *multi-carrier transmission technique*, which allows the transmission of data on multiple adjacent subcarriers, each subcarrier being modulated in a traditional manner with a linear modulation scheme such as QAM. In OFDM, the data for transmission is, via a serial to parallel converter, converted into several parallel streams, and each stream used to modulate a separate subcarrier. Thus, only a small amount of the total data is transmitted via each subcarrier, in a subchannel a fraction of the width of the total channel. As a result, in a multipath fading environment, as a fade notch moves across the channel, the fading appears to each subchannel almost as a flat fade. Figure 7.1 demonstrates such a scenario. The fade thus induces a significantly reduced amount of *intersymbol interference* (ISI) compared to that which would be experienced by a single carrier modulated system with a spectrum extending across the entire band. Further, while those subchannels at or close to the notch may experience a deep fade and hence thermal noise and ISI-induced burst errors, those removed from the notch won't. In the reconstructed original data stream, these burst errors are randomized, due to the interleaving which results from the parallel to serial process and therefore more easily corrected with FEC.

The robustness of OFDM to multipath interference is one of its most important properties. Figure 7.2 shows the spectrum of a standard four-channel *frequency division multiplex* (FDM) system. In such a scheme, modulated signals are stacked adjacent to each other with a guard band between each adjacent spectrum to ensure that there is no overlap between signals and to facilitate recovery via filtering of each signal at the receive end. While this approach works well, it suffers from the major drawback that, because of the spacing required between subcarriers, it wastes spectrum. As a result, it requires more bandwidth than would be required by a single carrier modulated by the original data stream, assuming that the same modulation is applied to the single carrier as to the subcarriers. With OFDM, however, the subcarriers are cleverly stacked close to each other. This results in overlapping spectra which (1) eliminates the spectral utilization drawback without incurring an adjacent inter-subcarrier interference penalty and (2) retains advantages in the multipath arena that accrue to parallel transmission of lower data rate streams. In general, for the same basic modulation method and same data rate, OFDM leads to better bandwidth efficiency and hence higher data capacity compared to standard FDM as well as single carrier transmission. It achieves the latter because its spectrum below the

Fig. 7.1 Multipath fade across a multi-carrier signal

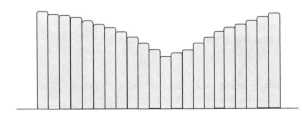

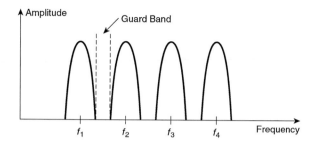

Fig. 7.2 Standard FDM frequency spectrum

main lobe of its first subcarrier and that above the main lobe of its last subcarrier falls off faster than that of a single carrier.

OFDM achieves its close stacking property, without adjacent channel interference, by making the individual subcarrier frequencies *orthogonal* to each other (more on orthogonality later). This is accomplished by having each subcarrier frequency be an integer multiple of the symbol rate of the modulating symbols and each subcarrier separated from its nearest neighbor(s) by the symbol rate. Thus, the multiple to generate each carrier is one integer different from those to generate its adjacent neighbors. Figure 7.3a shows an example of four subcarriers in the time domain over one symbol period, τ. Figure 7.3b shows these subcarriers in the frequency domain when each is modulated with symbols of period τ in the form of rectangular pulses. With such modulation, each subcarrier spectrum has the familiar sin x/x format, but note that, by the choice of subcarrier frequencies, the spectra overlap and each spectrum has a null at the center frequency of each of the other spectra in the system. Why this structure allows the individual modulated subcarriers to be demodulated with no interference from its neighbors is not intuitively obvious, at least not to the author, by simply looking at Fig. 7.3b. Remember, we are looking at a frequency domain representation, not a time domain representation of overlapping pulses. If anything, this figure seems to represent the ultimate in adjacent channel interference. The explanation here, like the devil, is in the detail, the detail being orthogonality.

The term orthogonal refers to the total uncorrelation between variables. In the case of OFDM the term orthogonal is used in reference to the mathematical relationship between subcarriers. For an OFDM system where the individual subcarrier modulating symbol period is τ, the symbol rate equals $1/\tau$. A cosine function derived subcarrier, s_n, of frequency n times the symbol rate is thus given by

$$s_n = \cos\left(2\pi nt / \tau\right) \tag{7.1}$$

and, when multiplied with itself and integrated over the period τ, we get

$$
\begin{aligned}
\int_0^\tau s_n \cdot s_n dt &= \int_0^\tau \cos\left(2\pi nt / \tau\right) \cdot \cos\left(2\pi nt / \tau\right) dt \\
&= \int_0^\tau \frac{1}{2}\left(1 + \cos\left(4\pi nt / \tau\right)\right) dt \\
&= \frac{\tau}{2}
\end{aligned}
\tag{7.2}
$$

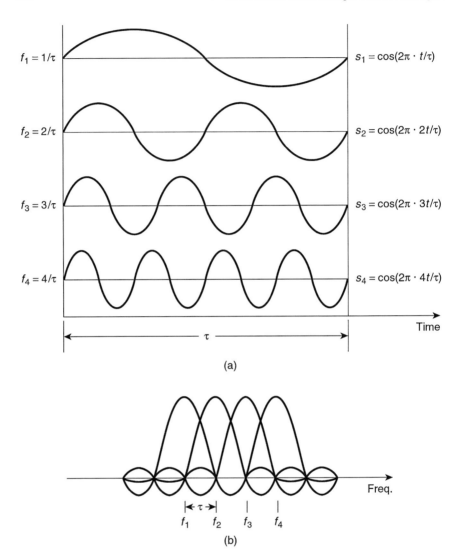

Fig. 7.3 Time and frequency representations of OFDM subcarriers. (**a**) Time domain representation of 4 OFDM subcarriers. (**b**) Frequency domain representation of 4 OFDM modulated subcarriers

(the latter term within the integral sign integrates to zero as the integration is over a whole number of cycles)

When, however, two different subcarriers, of frequencies n and m times the symbol rate, are cross multiplied and integrated over the period τ, we get

$$\int_0^\tau s_n \cdot s_m \, dt = \int_0^\tau \cos\left(2\pi nt / \tau\right) \cdot \cos\left(2\pi mt / \tau\right) dt$$

$$= \int_0^\tau \frac{1}{2}\left(\cos\left(2\pi\left(n+m\right)t / \tau\right)\right) + \cos\left(2\pi\left(n-m\right)t / \tau\right)\right) dt \qquad (7.3)$$

$$= 0$$

This latter result is because we are integrating both sinusoidal functions within the integral sign over a whole number of cycles. This is a direct consequence of our choice of subcarrier frequencies relative to the symbol rate $1/\tau$, and it is this relationship between cosine function subcarriers that make them orthogonal.

It can similarly be shown that with sin function derived subcarriers

$$\int_0^\tau \sin\left(2\pi nt / \tau\right) \cdot \sin\left(2\pi mt / \tau\right) dt = \frac{\tau}{2} \quad (m = n) \qquad (7.4a)$$

$$= 0 \quad (m \neq n) \qquad (7.4b)$$

and, further, that

$$\int_0^\tau \cos\left(2\pi nt / \tau\right) \cdot \sin\left(2\pi mt / \tau\right) dt = 0 \quad \text{(for all } n \text{ and } m) \qquad (7.5)$$

To see how OFDM takes advantage of these orthogonal properties in the demodulation process, consider an OFDM system with N subcarriers, with subcarrier frequencies varying from 0 times the symbol rate to N-1 times the symbol rate. Further, assume QAM modulated subcarriers, with the modulated subcarrier of frequency n times the symbol rate being given by

$$s_{QAM,n} = a_n \cos\left(2\pi nt / \tau\right) + b_n \sin\left(2\pi nt / \tau\right) \qquad (7.6)$$

Then the total OFDM signal, s_{OFDM}, is the sum of all such subcarriers and given by

$$S_{OFDM} = \sum_{n=0}^{n=N-1} s_{QAM,n} \qquad (7.7a)$$

$$= \sum_{n=0}^{n=N-1} \left\{ a_n \cos\left(2\pi nt / \tau\right) + b_n \sin\left(2\pi nt / \tau\right) \right\} \qquad (7.7b)$$

At the receiver, to decipher the symbol information a_k, S_{OFDM} is simply multiplied by $\cos\left(2\pi kt / \tau\right)$ and the product integrated over the period τ. This follows since

$$\int_0^\tau S_{OFDM} \cdot \cos\left(2\pi k t / \tau\right) dt = \sum_{n=0}^{n=N-1} \left\{ a_n \int_0^\tau \cos\left(2\pi n t / \tau\right) \cdot \cos\left(2\pi k t / \tau\right) dt \right.$$
$$\left. + b_n \int_0^\tau \sin\left(2\pi n t / \tau\right) \cdot \cos\left(2\pi k t / \tau\right) dt \right\} \tag{7.8a}$$

and by applying Eqs. (7.2), (7.3) and (7.5) to Eq. (7.8a), we get

$$\int_0^\tau S_{OFDM} \cdot \cos\left(2\pi k t / \tau\right) dt = \frac{\tau}{2} a_k \tag{7.8b}$$

Similarly, to decipher the symbol information b_k, S_{OFDM} is multiplied by sin $(2\pi k t / \tau)$ and the product integrated over the period τ, since

$$\int_0^\tau S_{OFDM} \cdot \sin\left(2\pi k t / \tau\right) dt = \sum_{n=0}^{n=N-1} \left\{ a_n \int_0^\tau \cos\left(2\pi n t / \tau\right) \cdot \sin\left(2\pi k t / \tau\right) dt \right.$$
$$\left. + b_n \int_0^\tau \sin\left(2\pi n t / \tau\right) \cdot \sin\left(2\pi k t / \tau\right) dt \right\} \tag{7.9a}$$

and by applying Eq. (7.4a), (7.4b) and (7.5) to Eq. (7.9a), we get

$$\int_0^\tau S_{OFDM} \cdot \sin\left(2\pi k t / \tau\right) dt = \frac{\tau}{2} b_k \tag{7.9b}$$

It is the orthogonal properties of the cos and sin functions stated earlier that generates these outcomes. Importantly, we note that the symbol deciphering process, in addition to rejecting all symbols than the desired one, is an "integrate and dump" one (Sect. 5.2.2), where the symbols are of rectangular shape and of equiprobable equal positive and negative values. This results in optimum error performance in the presence of thermal noise.

The OFDM signal S_{OFDM} is referred to as the baseband OFDM signal and in real systems is upconverted in the transmitter to the desired transmission frequency band. If the subcarriers are upconverted by f_c Hz, then the RF OFDM signal, $S_{OFDM,RF}$, is given by

$$S_{OFDM,RF} = \sum_{n=0}^{n=N-1} \left\{ a_n \cos\left(2\pi \left(f_c + n / \tau\right) t\right) + b_n \sin\left(2\pi \left(f_c + n / \tau\right) t\right) \right\} \tag{7.10}$$

In the receiver, the received signal is downconverter back to baseband prior to demodulation.

In order to avoid the construction of a large number of subchannel modulators in the transmitter and an equal number of demodulators in the receiver, modern OFDM systems utilize *digital signal processing* (DSP) devices. In fact, it's the availability of such devices that have made the commercialization of OFDM possible. Directly as a consequence of the orthogonality of the OFDM signal structure, modulation is able to be performed, in part, by using DSP to carry out an *inverse discrete Fourier transform* (IDFT). Similarly, demodulation is able to be performed, in part, by using DSP to carry out a *discrete Fourier transform* (DFT). The Fourier transform allows events in the time domain to be related to events in the frequency domain and vice versa for the inverse Fourier transform. The conventional transform, as discussed in Sect. 3.2.1, relates to continuous signals. However, digital signal processing is based on signal samples and so uses IDFT and DFT, which is a variant of the conventional transform and was reviewed in Sect. 3.2.2. In fact, it is typically the inverse *fast Fourier transform* (IFFT) and the *fast Fourier transform* (FFT) that are normally applied, these being a rapid mathematical method for computer applications of IDFT and DFT, respectively.

Figure 7.4 shows the basic processes in an IFFT/FFT-based OFDM system. The incoming serial data is first converted from serial to parallel in the S/P converter. If there are N subcarriers, N sets of parallel data streams are created. Each set contains a subset of parallel data streams, depending on the type of modulation. For example, if the modulation is 16-QAM, then each set contains four parallel data streams, the four bits in each symbol period of these streams being used to define a specific point in the 16-QAM constellation. The parallel data streams feed the mapper. For each subcarrier, the input data per symbol period is mapped into the complex number representing the amplitude and phase value of the subcarrier. For example, if the modulation is 16-QAM and the constellation diagram is as shown in Fig. 5.18b, the 1110 is mapped to the complex number $1 + j3$. The outputs of the mapper feed the IFFT processor. The IFFT knows the unmodulated subcarrier frequency associated with each input and uses this, along with the input, to fully define the modulated

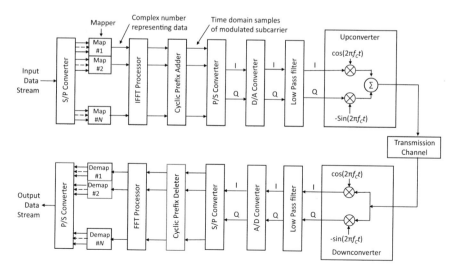

Fig. 7.4 Basic processes in IFFT/FFT based OFDM system

signal in the frequency domain. An IFFT is performed on this frequency representation, and the output is a set of time domain samples. The next process shown, the addition of a cyclic prefix, is optional, but almost always applied, and will be discussed below. The outputs of the cyclic prefix adder are fed to the P/S converter where they are broken down into their in-phase (I) and quadrature (Q) components and fed out serially in time to create a burst of samples per symbol. Note that the addition of the cyclic prefix can alternately be done on the signal after the P/S converter. The burst of serial samples is then transformed from discrete to analog (continuous) format via the digital-to-analog (D/A) converter and low-pass filter. The output of the low-pass filter is the baseband OFDM signal and is upconverted as shown to create the RF OFDM signal. The upconversion process shown results in a direct transfer of the baseband OFDM signal to RF. At the receiver, the process is reversed. The received waveform is first downconverted directly to baseband, low pass filtered, then digitized via an analog-to-digital (A/D) converter. The output of the A/D converter is feed to the S/P converter whose output then feeds the cyclic prefix deleter, should a cyclic prefix have been added in the transmitter. Alternatively, the cyclic prefix deleter can be located immediately following the A/D converter. The next step following the cyclic prefix deleter is conversion back to the complex representation of the symbols by the FFT. These complex representations are then de-mapped to recreate the original parallel data streams, which are then transformed to the original serial stream by the P/S converter. We note that the FFT demodulation process as a part of the demultiplexing is a discrete form of the "integrate and dump" continuous process discussed in Sect. 5.2.2.

We now take a closer look at OFDM symbol creation via the IFFT. Let's assume N subcarriers, each of frequency kf_0, where k varies from 0 to $N-1$. Assume further that each subcarrier is modulated by a complex symbol stream $X(k)$, where k varies from 0 to $N-1$, and where symbol rate $f_0 = 1/\tau$. Then each modulated subcarrier can be represented as $X(k)e^{j2\pi kf_0 t}$, and the composite signal $x(t)$ as shown in Fig. 7.5 is given by

$$x(t) = \sum_{k=0}^{N-1} X(k)e^{j2\pi kf_0 t} \tag{7.11}$$

The signal $x(t)$ represents an OFDM signal as it consists of equally spaced carriers of frequencies that are integer multiples of modulating symbol rate. To create one OFDM symbol in the discrete domain, suppose we sample $x(t)$ every T seconds, a total of N times, for an effective sampling time of NT and make $NT = \tau = 1/f_0$. Then in every symbol period τ, we create N discrete samples which together fully characterizes the N modulated subcarriers. The N discrete samples are taken at times $t = nT$, where n varies from 0 to $N-1$. If $x(n)$ is the sample of $x(t)$ at time $t = nT$, then, by substituting $t = nT$ and $f_0 = 1/NT$ into the right hand side of Eq. 7.11 we get

$$x(n) = \sum_{k=0}^{N-1} X(k)e^{\frac{j2\pi nk}{N}} \tag{7.12}$$

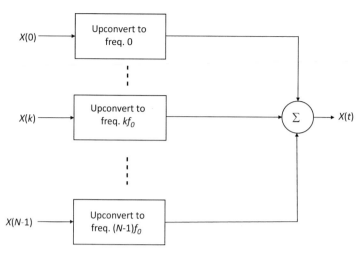

Fig. 7.5 OFDM symbol creation

What's special about the last equation? Yes, it's the IDFT of $X(k)$ where k varies from 0 to $N - 1$ as defined in Eq. 3.11! It's missing the $1/N$ multiplier, but as this is a constant, it's of no significance. This is why we can create an OFDM symbol by using an IDFT processor. In reality we use the more efficient IFFT processor. The output of the IFFT processor is a parallel stream of samples $x(n)$ of the multi-carrier signal. However, to create a real OFDM symbol, these samples must be fed out in order, spread T seconds apart. This why we have to next go through the parallel-to-serial converter process.

Figure 7.6 shows a simple OFDM baseband transmitter where the $N = 4$ and the individual subcarrier modulation is QPSK. The input data is transformed by the serial to parallel converter into four sets of data streams, each set consisting of two streams so that at any instant one of the our data combinations 00, 01, 10, or 11 can be sent to the adjoining map. The output of each map is a complex symbol $X(k)$. The maps feed the IFFT processor which creates four samples $x(k)$, where $k = 0,1,2,$ and 3. The P/S converter then spreads these samples over time as shown.

Figure 7.7 shows a very simple OFDM baseband transmitter and receiver that for simplicity excludes the transmitter parallel to serial converter and the receiver serial to parallel converter. For the system shown $N = 2$ and the individual subcarrier modulation is QPSK. The equations in the figure show how a short data stream is processed. The first two equations show how in the transmitter the map outputs $X(0)$ and $X(1)$ are converted by the IFFT to samples $x(0)$ and $x(1)$. The last two equations show how in the demodulator $x(0)$ and $x(1)$ are converted by the FFT to $\hat{X}(0)$ and $\hat{X}(1)$, estimates of $X(0)$ and $X(1)$.

Now that we have a fuller understanding of how the actual OFDM IFFT/FFT process, let's return to RF up- and downconversion that was briefly mentioned above. Here it is important to note that if on the transmit side, for a particular

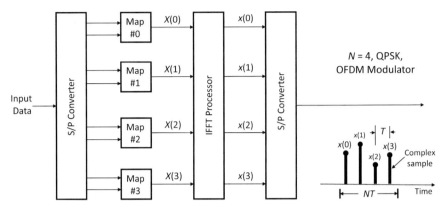

Fig. 7.6 Simple OFDM transmitter

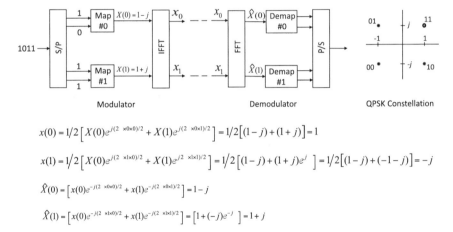

$$x(0) = 1/2\left[X(0)e^{j(2\pi\times0\times0)/2} + X(1)e^{j(2\pi\times0\times1)/2}\right] = 1/2\left[(1-j)+(1+j)\right] = 1$$

$$x(1) = 1/2\left[X(0)e^{j(2\pi\times1\times0)/2} + X(1)e^{j(2\pi\times1\times1)/2}\right] = 1/2\left[(1-j)+(1+j)e^{j\pi}\right] = 1/2\left[(1-j)+(-1-j)\right] = -j$$

$$\hat{X}(0) = \left[x(0)e^{-j(2\pi\times0\times0)/2} + x(1)e^{-j(2\pi\times0\times1)/2}\right] = 1-j$$

$$\hat{X}(1) = \left[x(0)e^{-j(2\pi\times1\times0)/2} + x(1)e^{-j(2\pi\times1\times1)/2}\right] = \left[1+(-j)e^{-j\pi}\right] = 1+j$$

Fig. 7.7 Data stream processing in a very simple OFDM transmitter and receiver

subcarrier, the QAM modulating complex symbol is $(a + jb)$, and the baseband subcarrier frequency is ω_b, then what is being upconverted is $(a + jb)e^{j\omega_b t}$, not $a\cos\omega_b t + b\sin\omega_b t$. The magnitude of the real part, I_M say, of $(a + jb)e^{j\omega_b t}$ is $[a\cos\omega_b t - b\sin\omega_b t]$, and the magnitude of the imaginary part, Q_M say, is $[b\cos\omega_b t + a\sin\omega_b t]$. To upconvert to an RF frequency ω_c, it is I_M that is multiplied with $\cos\omega_c t$, not a, and Q_M that is multiplied with $-\sin\omega_c t$, not b, and these two products when added create an RF subcarrier of the form $a\cos(\omega_c + \omega_b) - b\sin(\omega_c + \omega_b)$. At the receiver, this subcarrier, when downconverted via multiplication with $\cos\omega_c t$ and $-\sin\omega_c t$, results in scaled versions of I_M and Q_M.

Were the baseband symbol stream $x(n)$ as given by Eq. 7.12 to be upconverted by a carrier of frequency f_c, the resulting signal $s(t)$ say would be given by

$$s(t) = \sum_{k=0}^{N-1} X(k) e^{j2\pi(f_c + kf_0)t} \tag{7.13}$$

This would represent an RF signal where the lowest numbered subcarrier is at the carrier frequency and higher numbered subcarriers are all above the carrier frequency. In practice, however, we desire the carrier frequency to be at or very close to the center of the transmitted spectrum. This is accomplished by modifying the IFFT such that baseband sample stream is now $x'(n)$ say, given by

$$x'(n) = \sum_{k=0}^{N-1} X(k) e^{\frac{j2\pi n(k-k_c)}{N}} \tag{7.14}$$

where k_c is the subcarrier number corresponding to the center frequency of the rf signal. This subcarrier corresponds to the zero frequency out of the IFFT and is thus referred to as the *DC subcarrier*. For any $k < k_c$, the subcarrier is below the center frequency, and for any $k > k_c$, the subcarrier is above the center frequency.

As indicated above, OFDM is very robust in the face of multipath fading. Nonetheless, in the presence of such fading, a certain amount of ISI is unavoidable unless techniques are implemented to avoid it. Figure 7.8a is an illustration of how ISI can be incurred as a result of a delayed signal. One technique for eliminating, if not significantly reducing ISI, is the adding of a *guard interval*, or *cyclic prefix* (CP), of length τ_g to the beginning of each subcarrier transmitted useful symbol, of length τ_u, as shown in Fig. 7.8b. This prefix addition allows time for the multipath signals from the previous symbol to die away before the information from the current symbol is processed over the un-extended symbol period. Figure 7.8c is an illustration of how the ISI, shown in Fig. 7.8a, is avoided by using a cyclic prefix. Cyclic prefix addition is carried out, while the symbol is still in the form of IFFT samples and is achieved by copying the last section of the symbol, typically 1/16 to 1/4 of it, and adding it to the front. With this addition, the symbol total duration is now $\tau_t = \tau_g + \tau_u$. Due to the periodic nature of the modulated subcarrier, the junction between the prefix and the start of the original burst is continuous (more on this below). By adding the guard interval in this manner, the length of the symbol is extended while maintaining orthogonality between subcarriers. As long as the delayed signals from the previous symbol stay within the guard interval, then in the time τ_u there will be no ISI. Nonetheless, the symbol will likely experience a change to its amplitude and a shift of its phase as a result of the multipath effect created by delayed versions of itself. In the guard interval, there will be ISI, but the guard period is eliminated in the receive process, and the received symbol is processed only over the period τ_u. Should the delayed version of the previous symbol extend beyond the guard interval, however, ISI occurs, but is likely to be limited, as the strength of the delayed signals beyond the guard interval is likely to be small relative to the desired symbol. While adding a prefix eliminates or minimizes ISI, it is not without penalty, as it reduces data throughput, since N symbols are now transmitted over the period τ_t instead of over the shorter period τ_u. For this reason, τ_g is usually limited to no more than about 1/4 of τ_u.

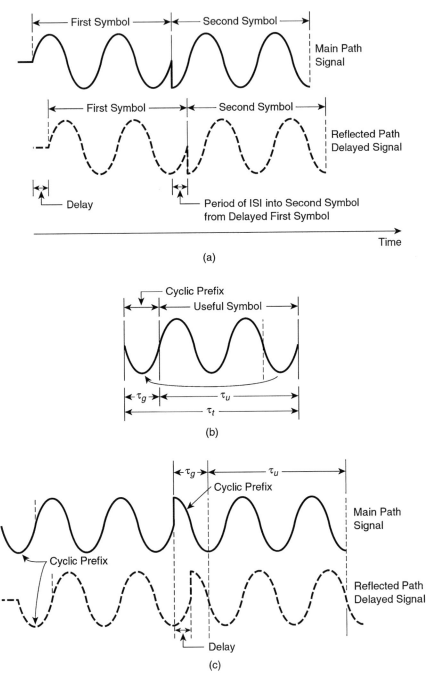

Fig. 7.8 ISI generation and elimination. (**a**) ISI due to delayed signal. (**b**) Cyclic prefix addition. (**c**) Elimination of ISI with cyclic prefix

The added prefix is called a cyclic prefix because it's created in such a way as to be a periodic extension of useful symbol. Recall that each subcarrier frequency is an integral multiple of useful symbol frequency. Thus, over the period of a useful symbol every subcarrier will contain an integral number of cycles. Because of this periodic nature, a section of the useful symbol copied from the front and added to the end results in no discontinuity at the junction, creating a continuous sinusoid. The symbol is thus extended in length while maintaining orthogonality. Figure 7.9 shows the creation of a cyclic prefix of length τ_g.

To create a cyclic prefix, complex symbols $X(k)$ out of the mapper are presented to the IFFT processor for a total of τ_t seconds. The IFFT processor uses this data to create a modulated symbol of length τ_u and then passes this symbol in form of time domain samples $x(n)$ to the cyclic prefix adder. Here a prefix is added to the received samples by taking copies of last samples that exist over time τ_g and adding them to front of the useful samples to create a symbol of length τ_t as shown, for example, in Fig. 7.10. We thus input data to IFFT processor, and, as is necessary, output data from the cyclic prefix adder at rate of $1/\tau_t$ symbols per second.

As mentioned above, multipath fading results in a change in amplitude and a phase rotation to each received OFDM symbol relative to the undistorted transmitted ensemble. For correct demodulation, these distortions must be corrected prior to de-mapping and parallel-to-serial conversion in receiver. This correction is achieved via the use of a so called one-tap (single-tap) equalizer. As indicated in Fig. 7.11, equalization is achieved, per subcarrier, by the complex multiplication of the distorted complex symbol output of the FFT processor by the inverse of the frequency response of the channel $H(f_i)$ at the corresponding subcarrier frequency [1]. $H(f_i)$ is normally determined by interpolating the channel response to pilot subcarriers located across the full transmission bandwidth.

As we saw in Chap. 6, decoding is improved by the use of soft bits out of the demodulator. For an OFDM system, this means soft bits out of the de-mapper. For a 4-QAM (QPSK) modulated two-bit sequence, the soft bit value of the first bit is simply a function of distance of a from y axis and of the second bit a function of the distance of b from x axis. However, to generate soft bit values for higher order QAM, a more complex algorithm must be applied to $a + jb$, as for example outlined in [2].

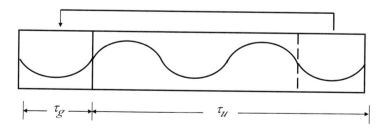

Fig. 7.9 Creation of a cyclic prefix in the continuous symbol domain

Fig. 7.10 Creation of a cyclic prefix in the discrete sample domain

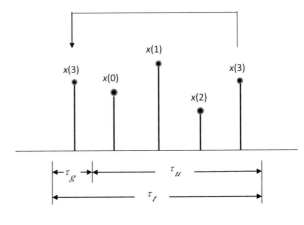

Fig. 7.11 OFDM one-tap equalization

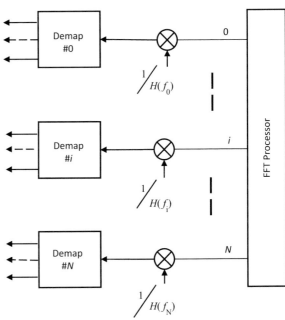

In real OFDM systems, the number of subcarriers actually created, N, is always less than the IFFT processor block size, N_{FFT}. The IFFT process is usually carried out on a total number of samples of size 2^x, where x is a positive integer. In real realizations, therefore, the IFFT processor block size, N_{FFT}, is chosen so that $N_{FFT} = 2^x \geq N$. For example, if number of real subcarriers is 200, then the smallest processor block size or number of "points" would be 256. There would thus, in this example, be 56 "null" carriers. This OFDM signal would be designated as a "256-point FFT" one, even though there would be only 200 real subcarriers.

As noted earlier, the IFFT processor samples the time domain signal $x(t)$ every T seconds. Also, $N_{FFT}T = \tau_u$. Thus, the *sampling rate* $F_S = 1/T = N_{FFT}/\tau_u$. This sampling rate is often specified in OFDM systems. In the frequency domain, it represents the bandwidth occupied by the main lobes of all subcarriers including null subcarriers. A null subcarrier is a bit of a misnomer as it means a subcarrier location where there is no subcarrier. When present, null subcarriers are usually placed symmetrically above and below the real subcarriers. In addition to being used to carry data, a few subcarriers are often used as pilot subcarriers as alluded to above. Pilot subcarriers are typically used for various synchronization and channel estimation purposes.

7.2.2 Peak-to-Average Power Ratio

A significant problem with implementation of OFDM is that it exhibits a high *peak to average power ratio* (PAPR), where PAPR is defined as the square of the peak amplitude divided by the mean power. The reason for the high PAPR is that in the time domain, an ODFM signal is the sum of N separate subcarriers which are modulated sinusoidal signals. The amplitudes and phases of these sinusoids are uncorrelated, but in the normal course of operation certain input data sequences occur that cause all the sinusoids to add in phase leading to a signal with a very high peak relative to the average. It can be shown [3] that the probability that the PAPR of an OFDM signal is above a certain threshold, $PAPR_0$ say, is given by

$$\Pr\left(PAPR > PAPR_0\right) = 1 - \left(1 - e^{-PAPR_0}\right)^N \tag{7.15}$$

where N is the number of subcarriers.

Figure 7.12 shows plots of the distribution of the PAPR given by Eq. 7.15 for different values of the number of subcarriers N. The figure shows that for a given threshold $PAPR_0$, the probability that the PAPR exceeds that threshold increases with N. Note, however, that this formula implies that the PAPR is independent of modulation order. Numerous computer simulations confirm this.

The input-output characteristics of a power amplifiers exhibit a linear range above which nonlinearity sets in until ultimately the output level reaches a maximum. If a high PAPR signal is transmitted through a power amplifier and the peaks of signal fall in the nonlinear region, then signal distortion occurs. This distortion shows up as intermodulation among the subcarriers and out of band emissions. The net result of OFDM signals having a high PAPR is that the power amplifier has to be operated with a large power back off leading to inefficient operation. In addition to potential problems with the transmitter power amplifier, a high PAPR requires a large dynamic range in the transmitter D/A converter and the receiver A/D converter.

A number of techniques have been developed to minimize the PAPR of OFDM systems, including clipping which is the simplest such technique. With clipping, all signal amplitudes above a predetermined threshold are clipped to the threshold value. Clipping reduces the PAPR at the expense of both in-band and out-of-band distortion and so must be applied judiciously. The in-band distortion results in an

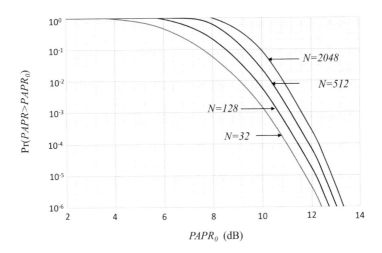

Fig. 7.12 PAPR distribution for different numbers of OFDM subcarriers

increase in the BER whereas the out-of-band distortion manifests itself as spectral spreading. Other OFDM PAPR reduction techniques will not be reviewed here but can be found in [4].

7.2.3 Frequency and Timing Synchronization

In order to successfully demodulate an OFDM signal, the receiver needs to perform three major synchronizations:

Carrier frequency synchronization, which is the process of reducing or eliminating *carrier frequency offset* (CFO). This offset is the difference between the frequency received at the receiver and that of the receiver local oscillator and is caused by the difference between the transmitter and receiver RF local oscillators, the mobile channel Doppler shift, and oscillator instabilities. CFO results in a loss of orthogonality between the subcarriers which in turn leads to intercarrier interference (ICI). The degradation of BER caused by CFO is function of the modulation order. It is shown in [5] that with QPSK modulation OFDM can perform acceptably with a CFO of up to about 5% of the subcarrier spacing whereas, with 64-QAM, the CFO must be $\leq 1\%$.

Symbol timing synchronization, which is the process of reducing or eliminating the timing offset between where the receiver believes that each symbol into the FFT starts and the optimal time where the symbol should start. Ideally, this offset should be zero. However, should a cyclic prefix have been added, this requirement is relaxed somewhat. If fact, once the timing offset is positive and less than the difference between the cyclic prefix length and the channel impulse response, then there is no degradation in performance as the period over which the symbol is processed is free of ISI. If, however, the above condition is not met, then ISI and ICI occur.

Sampling clock synchronization, which is the process of reducing or eliminating the difference between the frequency/phase of the transmitter local oscillator used to generate the sampling clock and that of the receiver local oscillator used to generate the sampling clock. This difference can cause the receiver to assume an incorrect subcarrier spacing, resulting in a gradual increase in ICI with subcarrier number, and a gradual increase in symbol timing offset.

Synchronization methods can be data-aided or non-data-aided:

Data-aided synchronization: Here extra information is included in the transmitted data and may be in the form of:

- A *preamble*, i.e., OFDM *training symbols*, typically one or two symbols ahead of a fixed block of symbols containing control data and user data. These training symbols contain only subcarriers specially modulated to facilitate synchronization.
- *Pilot symbols*, i.e., specially modulated subcarriers interspersed repetitively among control and user data modulated subcarriers. At the receiver the general structure of the synchronization modulating data is known, and the received data is compared against the known data or earlier received data and correlation between the two used to attain synchronization.

Non data-aided synchronization or *blind synchronization*. Here:

- Receiver typically uses cyclic prefix to achieve synchronization. An OFDM symbol created from N modulated subcarriers generates N samples in time domain. If last L samples copied and used as cyclic prefix, then for every OFDM symbol, there are L pairs of identical samples separated by time period of N samples, τ_u, the inverse of which is the fundamental subcarrier frequency. This sample redundancy is used to estimate both timing and frequency offsets. However, this approach can only estimate the CFO if it is within the range of ± 0.5 of the subcarrier spacing.

5G NR specifies two sequences, broadcast from each base station, to permit the receiving mobile unit to achieve time and frequency synchronization and acquire other useful parameters such as the cyclic prefix. These sequences are the *primary synchronization sequence* (PSS) and the *secondary synchronization sequence* (SSS). The UE uses the PSS, SSS, and specific synchronization algorithms to estimate and adjust the time and frequency offsets. A description of the functioning of these sequences can be found in Chap. 10 and in [6].

7.2.4   Phase Noise Effect on Performance

In an ideal oscillator, the signal created is a pure sine wave and thus with no variation in frequency over time. In a real oscillator, this is not the case. Here signals suffer disturbances due to thermal noise and device instability. These disturbances have little impact on the signal amplitude close to center frequency but show up mostly as a random variation of the phase and hence frequency about the center position. This phenomenon is referred to as *phase noise* (PN). In OFDM systems phase noise seen at the receiver is the net result of that due to the transmitter upconverter oscillator and the receiver downconverter oscillator. OFDM is more sensitive to PN than single carrier systems because here the PN results in non-optimum random frequency sampling of the subcarriers which in turn destroys orthogonality between the subcarriers and hence produces ICI.

The characteristics of an oscillator's phase noise are usually defined via its single-sided power spectral density (PSD). There are a number of models for oscillator PSD, and predicted PSD about the oscillator center frequency tends to vary by model. A simple but enduring model is that proposed by Leeson in 1966 [7] for linear feedback free-running oscillators. Leeson's model predicts:

- PSD directly proportional to the noise figure of the oscillator's buffer amplifier and to the oscillator temperature and inversely proportional to signal strength
- PSD approximately proportional to the square of the oscillator frequency, i.e., increasing by approximately 6 dB for every doubling of the oscillator frequency
- An initial decrease of about 30 dB per decade up to the point where $1/f$ noise effects no longer predominate
- Changing from that point to about 20 dB per decade up to the feedback loop half bandwidth
- Flattening out thereafter

A graphical presentation based on this model is shown in Fig. 7.13 [8]. In the model by Demir et al. [9] proposed in 2000, the PSD of a free-running oscillator is modelled as being flat over a range close to its center frequency then falling off at a rate of 20 dB per decade. A more recent model [8] is for phase-locked loop (PLL) oscillators which tend to exhibit lower PN than free running ones but are more costly to implement. This model shows an initial decrease of about 10 dB per decade changing to about 20 dB per decade then levelling off as shown in Fig. 7.14. All three models mentioned above predict that overall, phase noise increases by approximately 6 dB per doubling of the center frequency, and this has indeed found to be the case in practice.

PN creates two effects when viewed on a constellation diagram. One effect is a rotation of the constellation caused by what is referred to as *common phase error* (CPE), i.e., a phase error that is common to all subcarriers. Fortunately, CPE can be estimated from reference signals and can thus be removed. The second effect is fuzzy constellation points and is due to the ICI resulting from loss of orthogonality between the subcarriers. The latter is hard to eliminate but its impact can be reduced

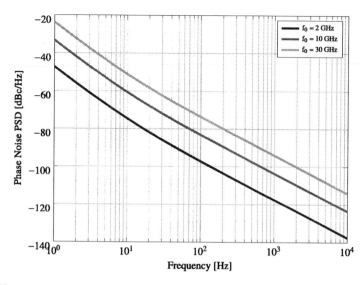

Fig. 7.13 Phase noise PSD as per Leeson. Lower, middle, and upper traces are 2, 10, and 30 GHz, respectively. (From [18], with permission of Elsevier)

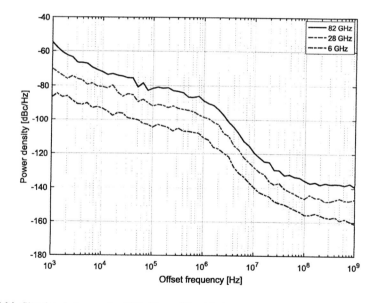

Fig. 7.14 Simulated phase noise PSD. (From [8], with the permission of Elsevier)

by increasing the subcarrier spacing. The influence of PN on the received signal in an OFDM system employing QPSK modulation and a free-running oscillator having a phase noise PSD 3 dB bandwidth of Δf_{3dB} is shown in the constellation diagrams of Fig. 7.15 [10]. The result shown in Fig. 7.15a is for the case where the oscillator corner frequency is low compared to the subcarrier spacing. As can be seen, the major effect is phase rotation, the result of CPE. The result shown in Fig. 7.15b is for the case where, for the same total oscillator power as in the first case, Δf_{3dB} is high compared to the subcarrier spacing. Here, rotational behavior is seen to be replaced by random fuzzy behavior about the true constellation points resulting from ICI. We note that the higher the modulation order, for example, 16-QAM versus QPSK, the more closely spaced are the points in the constellation diagram, and hence the more sensitive the system's BER performance to PN.

In 5G NR, increased data capacity is afforded by the use of mm wave frequencies. However, as indicated above, phase noise PSD increases with carrier frequency. Thus, PN is a much greater issue at mm wave frequencies than at sub-6 GHz ones and without adequate correction performance could be significantly degraded. 5G NR has options for wider subcarrier spacing than that afforded by LTE (up to 240 kHz versus 15 kHz). Thus, use of higher subcarrier spacing will reduce phase noise-induced ICI. To address CPE 5G NR has introduced the phase tracking reference signal (PT-RS) which can also be used for ICI reduction [11].

7.2.5 Transmitter Windowing and Filtering

The OFDM spectrum, beyond the section occupied by the main lobes of all subcarriers, decays somewhat slowly on both sides. The main reason for this somewhat slow decay is that the standard OFDM IFFT created symbol with added cyclic prefix has vertical boundaries creating a rectangular pulse and hence generating a sin x/x-shaped spectrum as we saw in Chap. 3. To meet stringent spectral masks stipulated by standardization bodies, action can be taken to minimize this spectral spread. This action can take the form of *windowing*, *filtering*, or both.

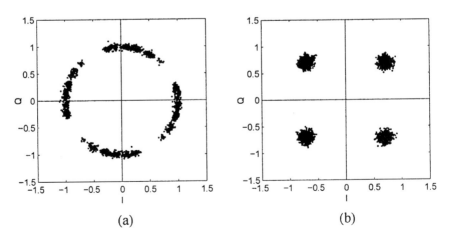

(a)   (b)

Fig. 7.15 Influence of PN on QPSK symbols. (**a**) Low Δf_{3dB}, (**b**) high Δf_{3dB}. (From [10], with the permission of the authors)

Windowing reduces spectral spread by changing the shape of the symbol edges to create a smoother transition. Figure 7.16 shows one way that this can be done. Here the start of the CP is changed from a vertical line to one with a linear slope, the line starting at zero and increasing to the maximum value and occupying a time period t_x. Further, a cyclic extension, also of time period t_x, and created by copying samples from the start of the useful period, is added to the end of the useful period, starting at the maximum value and decreasing linearly to zero. The overall symbol length is thus increased by t_x. The succeeding symbol starts at its normal start time. No delay has thus been added to the system. There is now, however, an overlap of symbols at the end of one and the start of the other lasting for t_x. These less abrupt start and finish symbol transitions lead to a more contained spectrum. For a delayed symbol stream, delayed by the full cyclic prefix time as shown in Fig. 7.16, the first t_x period of this symbol, which has been altered, is added to the non-delayed symbol and thus "interferes" with it, i.e., causing ISI.

In real systems, it is common to create windowed pulses by using the raised cosine shape. This is the same shape introduced in Chap. 3, but with the shape here being a function of time, not frequency. With raised cosine windowing we get a smooth transition out of one pulse and into the next and hence good spectral containment. This containment is not without trade-off, however. We have effectively decreased the cyclic prefix to $t_g - t_x$. Shown in Fig. 7.17 [8] is an OFDM spectrum with no windowing and one with windowing employing linear slopes. We note that the difference in spectral spread is significant.

As indicated above, another way of containing OFDM out of band emissions is by filtering. This is done to the entire signal at the output of the modulator. However, often the transmitted signal is contained in non-contiguous sub-bands, making filtering cumbersome as a separate filter is needed for each sub-band. Filter bandwidth is typically close to the size the transmission band or sub-band and designed so that only a few subcarriers close to the edges of the sub-band are affected. Nonetheless, BER performance is somewhat degraded; thus, as with windowing, there is a trade-off between spectral containment and system performance.

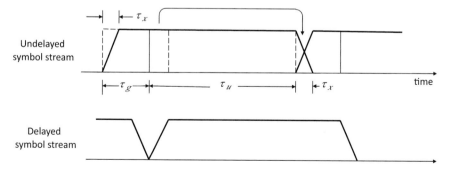

Fig. 7.16 OFDM transmitter windowing

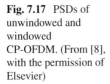

Fig. 7.17 PSDs of unwindowed and windowed CP-OFDM. (From [8], with the permission of Elsevier)

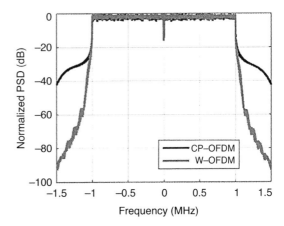

In 5G NR neither windowing nor filtering in the transmitter is specified. However, the implementer is free to apply any operation such as windowing/filtering to improve spectrum containment, so long as it is transparent to the receiver and the overall system meets stated specifications.

7.3 Orthogonal Frequency Division Multiple Access (OFDMA)

OFDM is a multi-carrier transmission technique that enables transmission between two points and where all subcarriers are assigned to one user. However, it lends itself to form the basis of a multiaccess technique for both DL and UL transmission, which is referred to as *orthogonal frequency division multiple access (OFDMA)*. It is used in 5G NR in both the downlink and uplink and here is often referred to as *cyclic prefix OFDM* (CP-OFDM). In OFDMA, OFDM subcarriers are divided into sets called *subchannels* and these subchannels assigned to different users in a given time slot, i.e., in a given a number of consecutive OFDM symbols. Subcarriers in each subchannel can be either distributed, i.e., spread over the full channel spectrum available or localized, i.e., structured adjacent to each other. Figure 7.18a, b depicts a three subchannel frequency assignment example where allocation is distributed and localized, respectively.

With distributed allocation, some subcarriers on a given link will likely experience good signal to interference and noise ratio (SINR), while others will not, thus experiencing high BER. However, interleaving and coding can minimize the errors generated in the low SINR subcarriers.

Since users typically have different locations, the individual BS/MU links of these users typically suffer from different multipath fading and thus different frequency response across the channel. Localized allocation attempts to exploit this phenomenon by assigning to each user a portion of the available spectrum where channel conditions are good, i.e., with a high SINR. In order to accomplish this, it

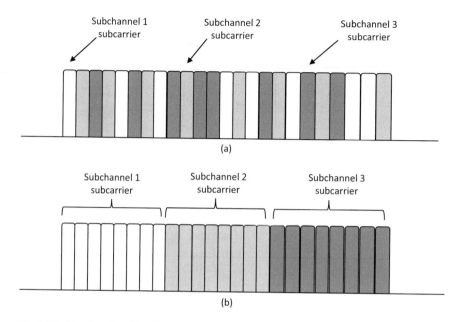

Fig. 7.18 Distributed and localized allocation example. (**a**) Distributed allocation. (**b**) Localized allocation

is obvious that accurate SINR data across the entire useable spectrum must be available to each receiver and that this information be communicated in a timely fashion to its associated transmitter.

If accurate SINR data can be derived at the receiver and applied quickly enough so that a good localized channel can be utilized before the channel condition changes, then localized allocation should outperform distributed allocation. Should this not be the case however, such as where the user is in a highly mobile state, then distributed allocation would be preferred.

With OFDMA as the DL scheme, subchannels are assigned by the *medium access control* (MAC) (Sect. 10.5.4), each subchannel addressed to a different MU, with modulation, coding, etc., tailored to each BS/MU link. With OFDMA as the UL access scheme, subchannels are assigned to users via MAC messages sent downstream by the BS, and several MU transmitters may transmit simultaneously with modulation, coding, etc., again tailored to each BS/MU link. Typically, transmit power at a MU is less than that at the BS, resulting in an imbalance in uplink and downlink system gains if the same number of subcarriers are transmitted in each direction. Subchanneling, however, concentrates MU transmit power in fewer subcarriers, enabling system gains to be similar for both up and down links. As an example, if data subcarriers are divided into 16 subchannels, a 12 dB system gain improvement is achieved in an uplink carrying one subchannel, albeit at the expense of lower data throughput. This system gain increase results from a 12-dB improvement in the BS receiver sensitivity.

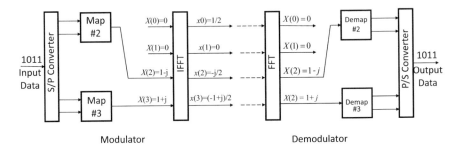

Modulator   Demodulator

Fig. 7.19 A simplified OFDMA baseband transmission scheme

Fig. 7.20 OFDMA
scheduling example

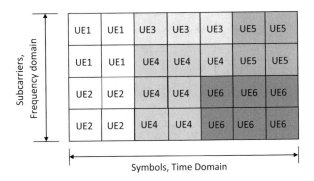

To get a sense of the IFFT and FFT processing associated with OFDMA transmission, a highly simplified OFDMA baseband transmission scheme is depicted in Fig. 7.19 where a user, User 1, is sending the data stream 1 0 1 1. Here the modulation is QPSK, the number of available subcarriers N is 4, and two such subcarriers are assigned to the user. Note, therefore, that at the IFFT processor two of the inputs are 0. A user at a second MU, User 2 say, is thus able utilize these two subcarriers simultaneously with User 1.

In OFDMA, the scheduler schedules on a two-dimensional (frequency x time) canvas taking into account the individual user capacity requirements and the channel condition across the full channel bandwidth. In Fig. 7.20 we see an example of how six different users may be scheduled in the time/frequency plane.

7.4 Discrete Fourier Transform Spread OFDM (DFTS-OFDM)

Discrete Fourier transform spread OFDM (DFTS-OFDM) is a modified version of OFDMA used optionally by 5G NR in the UL. As in OFDMA, the transmitted signal in a DFTS-OFDM system is a number of orthogonal subcarriers. Note, therefore, that though also referred to as single carrier-FDMA (SC-FDMA), the transmitted signal is not a single carrier! The SC-FDMA nomenclature comes because, in a specific circumstance, the PAPR of DFTS-OFDM is the same as a truly SC modulated signal, this PAPR being considerably less than the PAPR with standard OFDMA. However, even if the above-referenced specific circumstance is not met, the PAPR of DFTS-OFDM, though more than that of a truly SC modulated signal, is still less than that with standard OFDMA. It is this improved PAPR performance of DFTS-OFDM that drives its implementation.

As was shown earlier, OFDMA takes, per user, N data symbols say (a data symbol being a grouping of input bits), runs them through constellation mappers to create subcarriers in the form of complex numbers, then processes these complex numbers by an M-point ($M > N$) IDFT to generate time domain samples. Thus, within an OFDMA symbol, each subcarrier is modulated with one data symbol. DFTS-OFDM, in contrast, first feeds the N outputs of the mappers into an N-point DFT processor which creates subcarriers in the form of complex numbers and then processes these complex numbers by an M-point IDFT processor to generate time domain samples. The output of the DFT processor spreads input data symbols over all subcarriers, hence the nomenclature DFT-spread OFDM. The only physical difference between DFTS-OFDM as shown in Fig. 7.21 and OFDMA is that in DFTS-OFDM, a DFT processor has been added in the transmitter and an IDFT processor added in the receiver.

In DFTS-OFDM there are two defined categories of subcarrier mapping:

- *Localized subcarrier mapping*, referred to as localized FDMA (LFDMA). Here the DFT outputs are allocated to consecutive subcarriers with zeros occupying unused subcarrier positions.
- *Distributed subcarrier mapping*, referred to as distributed FDMA (DFDMA). Here the DFT outputs are allocated over the entire bandwidth with zeros occupying unused subcarrier positions.

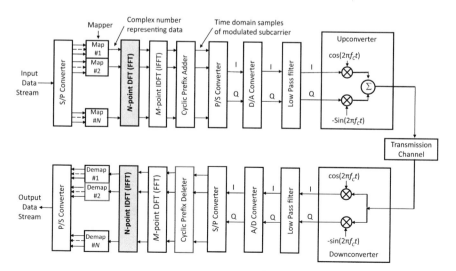

Fig. 7.21 Transmitter/receiver block diagram of DFTS-OFDM

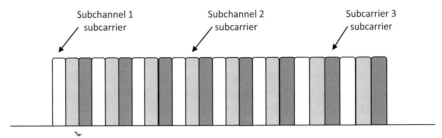

Fig. 7.22 DFTS-OFDM IFDMA subcarrier allocation

Interleaved FDMA (IFDMA) is the special case of DFDMA where $N = M/Q$, where Q, the bandwidth expansion factor, is a whole number, and where there is equidistance between the occupied subcarriers. For IFDMA, time symbols out of the M-point IDFT are simply a repetition of the original symbols into the DFT with a scaling factor of $1/Q$ and some possible phase rotation. But the PAPR of symbols into the DFT is the PAPR of the *single carrier* modulation whose constellation the mapper creates. Thus, as will be seen below, the average of the PAPR of the IFDMA signal is the same as that of the single carrier modulation whose constellation the mapper creates. IFDMA is the specific circumstance mentioned earlier that, in reference to PAPR at least, supports the nomenclature SC-FDMA. For subcarrier distributions other than IFDMA, the time signal is the exact copies of original input signals to DFT in some but not all M sample positions. As a result, the PAPR is higher than the underlying modulation, but less than with OFDMA and the same modulation. The disadvantage of IFDMA relative to LFDMA is that it provides less flexibility for scheduling. Shown in Fig. 7.22 is IFDMA mapping for the case where the number of subchannels is 3, $M = 24$ and $N = 8$.

To aid in understanding how the IFDMA allocation results in a lower PAPR compared to LDFMA allocation, consider the highly simplified DFTS-OFDM scenarios outlined in Fig. 7.23 for the case where $N = 2$, $M = 4$, the modulation is QPSK, and the input data stream is 1 0 1 1. For each scenario, the power associated with each time sample $y(x)$ out of the IDFT processor is given by $|y(x)|^2$. For the LFDMA scenario, the power associated with the four time samples computes to ½, 1, ½, 0, respectively. Thus, the peak power is 1, generated by $y(1)$, and the average power is ½, leading to a PAPR of 2 or 3 dB. What is most interesting, however, is that for the IFDMA scenario, the power associated with all four time samples are the same, namely, ½; thus the PAPR is 1 or 0 dB.

Shown in Fig. 7.24 [12] are graphs comparing the PAPR performance of IFDMA, LFDMA, and OFDMA for 4-QAM (QPSK), 16-QAM, and 64-QAM, where $M = 256$ and $N = 64$. These graphs show the peak-to-average power ratio $PAPR_0$, for which $PAPR_0$ is exceed a given fraction, X say, of the time. From these graphs the data in Table 7.1 can be obtained.

From the above results and the PAPR graphs of simulations in [12], the following is indicated:

- OFDMA PAPR is independent of modulation order, as note in Sect. 7.2.2.
- For IFDMA and LFDMA, for a given probability of occurrence, the PAPR increases with modulation order.

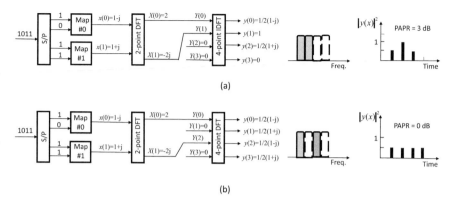

Fig. 7.23 LFDMA and IFDMA comparison. (a) LFDMA (b) IFDMA

- For IFDMA, where the probability that the PAPR > $PAPR_0$ is 50%, i.e., on average, $PAPR_0$, for the three modulations shown, is the same as that indicated in Sect. 5.3.7 for a single modulated carrier.
- For IFDMA and QPSK, the PAPR is always 0.

DFTS-OFDM's low PAPR compared to OFDMA makes it an attractive alternative to OFDMA in mobile UL communications. This is because power is a limited resource at the MU and a lower PAPR leads to a more power efficient output amplifier. For 5G NR, DFTS-OFDM is optional for the uplink, the other option being OFDMA with cyclic prefix.

As we have seen above, the PAPR when using DFTS-OFDM, unlike OFDMA, is a function of the underlying subcarrier modulation order, be it LDFMA or IDFMA. Thus, for a given modulation order, should the underlying subcarrier PAPR be lowered, then the PAPR of DFTS-OFDM would be lowered. Such lowering is possible for 64-QAM and 256-QAM should the standard signal point square constellation diagram be modified to be non-square as shown by Morais in [13]. For 64-QAM, where the PAPR of the square constellation is 3.7 dB, the proposed constellation, which has the general shape shown in Fig. 7.25, exhibits a PAPR of 2.50 dB, for a net reduction of 1.2 dB and hence, for a given power amplifier, an increase in maximum output power of 1.2 dB. For 256-QAM, where the PAPR of the square constellation is 4.2 dB, the proposed constellation exhibits a PAPR of 3.0 dB, for a net reduction also of 1.2 dB and hence, for a given power amplifier, an increase in maximum output power also of 1.2 dB.

For these non-square constellations, the bit error rate performances are slightly degraded relative to their associated square constellations. For the non-square 64-QAM constellation, the increase in required SNR for a BER of between 10^{-3} and 10^{-6} is approximately 0.3 dB. Thus, the net link margin improvement with its use would be approximately 1.2–0.3 = 0.9 dB. For the non-square 256-QAM constellation, the increase in required SNR for a BER of between 10^{-3} and 10^{-6} is

Fig. 7.24 PAPR
performance of IFDMA,
LFDMA, and
OFDMA. (From [12], with
the permission of the
International Journal of
Computer Applications)

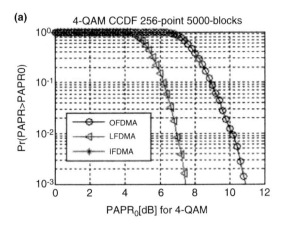

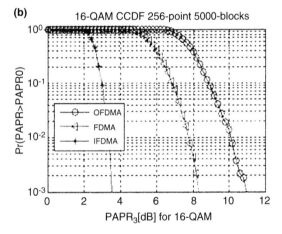

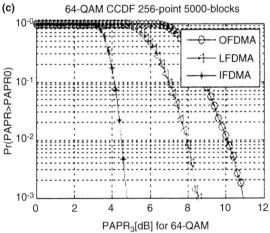

Table 7.1 PAPR (dB) performance of IFDMA, LFDMA, and OFDMA

Mapping	X	QPSK	16-QAM	64-QAM
IFDMA	0.5	0	2.6	3.7
LFDMA	0.5	5.4	5.7	6.0
OFDMA	0.5	7.7	7.7	7.7
IFDMA	0.001	0	3.5	4.8
LFDMA	0.001	7.5	8.3	8.6
OFDMA	0.001	10.9	10.9	10.9

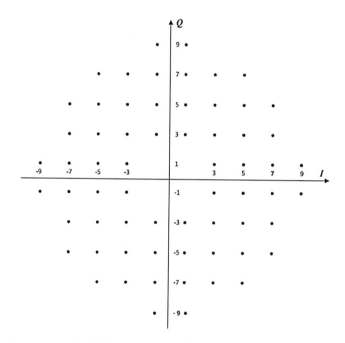

Fig. 7.25 Non-square 64-QAM signal point constellation

approximately 0.4 dB. Thus, the net link margin improvement with its use would be approximately $1.2 - 0.4 = 0.8$ dB. A method to perform hard and soft bit de-mapping for these non-square constellations is given in [14].

7.5   Other 5G NR Considered Waveforms

In moving from 4G to 5G, 3GPP considered several alternative waveforms that held the promise of improving on the performance OFDMA, often referred in 5G NR as cyclic prefix OFDM (CP-OFDM), when used as the underlying waveform. After much study and taking into account all the advantages and disadvantages offered by these alternative waveforms, including practical ramifications, 3GPP opted to remain with CP-OFDM, as it concluded that none of the alternatives presented a compelling case for change. That said, it is certainly possible, that given future improvements to these techniques, one or more could be considered in the future for application in 6G or beyond networks. In the following sections, three candidates that were strongly considered are addressed.

7.5.1   Filter Bank Multi-carrier (FBMC)

Filter bank multi-carrier (FBMC) [15] is an OFDM-derived waveform where subcarriers are individually filtered so as to suppress their side lobes resulting in reduced out of band emissions relative to OFDM. In OFDM, the "filtered" symbol is a rectangular pulse which, unfiltered, results in the well-known $\sin x/x$ frequency spectrum. As we saw in Chap. 5, if we contain the spectrum of a rectangular pulse by filtering in the frequency domain with a shape that meets Nyquist's criteria, we get a pulse as shown in Fig. 5.4 with a much longer time duration that overlaps with adjacent symbols, but of zero value at the center of these symbols. FBMC uses such filtering to contain spectrum.

In FBMC, the filtering of subcarriers is done in the baseband digital domain, right after the IFFT process. A filter is implemented for each subcarrier and all such filters aligned into a filter bank, hence the nomenclature of the waveform. The first filter of the bank, the filter associated with the zero-frequency subcarrier, is called the prototype filter. The filters for the other subcarriers are created by taking a copy of the prototype filter and shifting it to their individual frequencies. A filter bank so created is referred to as a polyphase filter bank. Prototype filters are characterized by their overlapping factor K, where K is the ratio of the duration of the filter's impulse response to the multi-carrier symbol period. The factor K is the number of multi-carrier symbols that overlap in the time domain. The prototype filters used with FBMC are typically root-raised cosine types as described in Chap. 5, and K is typically chosen to be 4. In Fig. 7.26a [8] we show the frequency response of an FBMC prototype filter with a factor $K = 4$ along with that created by a rectangular pulse as in OFDM. In Fig. 7.26b [8] we show the time response of this $K = 4$ prototype filter along with the OFDM rectangular time response.

Because of the spreading of individual symbols into adjacent symbols, the classic OFDM orthogonality is impacted, and thus FMBC is not orthogonal with respect to the complex plane. However, with tight filtering as with the prototype filter of Fig. 7.26, all even numbered subcarriers do not overlap with each other, are thus orthogonal to each other, and do not induce intercarrier interference (ICI) in each other. The same applies to all odd numbered subcarriers.

We are thus left with the issue of ICI between adjacent subcarriers. With FBMC we eliminate this problem via the use of offset quadrature amplitude modulation (OQAM). With OQAM, on a given subcarrier we offset the in-phase PAM symbols

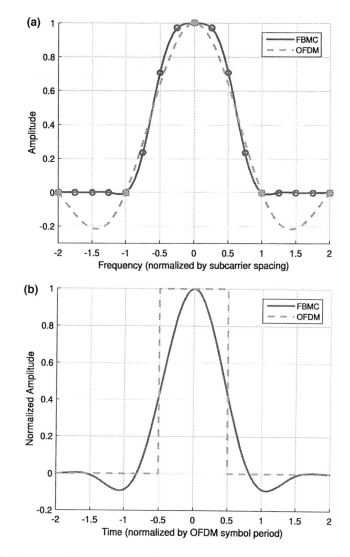

Fig. 7.26 Frequency and time responses of prototype filter of FMBC ($K = 4$) and OFDM. (From [8], with the permission of Elsevier)

by half the symbol length relative to the quadrature components. On its adjacent subcarriers, we offset the quadrature symbols by half the symbol length relative to the in-phase symbols. This offset concept is illustrated in Fig. 7.27 where we show in-phase and quadrature symbol placement on a time frequency plane.

As FBMC produces a well-localized subchannel in both the frequency and time domain, CP is not required. This removal of the CP results in improved spectral efficiency relative to OFDM. This is in addition to the gain in spectral efficiency as a result of the reduced band edges, resulting in turn from the subcarrier filtering.

Fig. 7.27 Offset QAM
symbol, placement

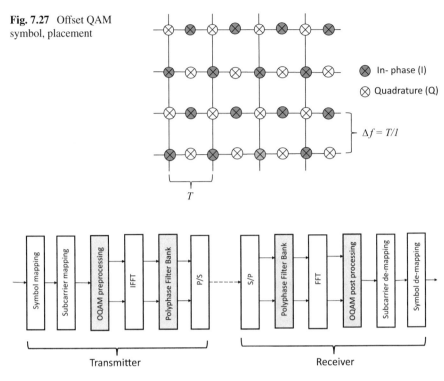

Fig. 7.28 FBMC transmitter/receiver block diagram

Shown in Fig. 7.28 is a block diagram of a FBMC baseband transmitter/receiver system. We note that in the receiver a polyphase filter bank is also required in order to achieve matched filtering, necessary for optimum signal detection.

One of the big promises of FBMC is lower *out-of-band* (OOB) *emissions* relative to CP-OFDM. And indeed, at low output power amplifier levels where amplification is essentially linear, this is the case. At higher powers, however, more typical of real systems, the advantage all but disappears. Results published in Ref. [16] indicate that when subject to a nonlinear power amplifier with input power of −4 dBm, and output power of 24 dBm, FBMC ($K = 4$) and CP-OFDM have almost the same OOB leakage.

7.5.2 Generalized Frequency Division Multiplexing (GFDM)

Generalized frequency division multiplexing (GFDM) [17] is a flexible, multi-carrier transmission technique which has some similarities to OFDM. Incoming modulation symbols are spread across two dimensions (time and frequency) by being processed block-by-block, where each block comprises K subcarriers and M "sub-symbols" and hence contains $N = MK$ modulation symbols. With OFDM K

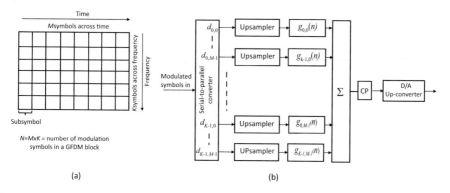

Fig. 7.29 GFDM symbol block structure and transmitter block diagram. (**a**) Symbol block structure. (**b**) Transmitter block diagram

modulation symbols are transmitted with K subcarriers in one time slot. With GFDM, however, MK modulation symbols are individually filtered and transmitted over K subcarriers in M time slots. The frequency spectrum of each subcarrier is therefore restricted, resulting in reduced out of band emissions and hence higher spectral efficiency relative to CP-OFDM. The structure of a block is shown in Fig. 7.29a. The block diagram of GFDM transmitter is shown in Fig. 7.29b. Encoded user data symbols are mapped into N complex valued QAM symbols which are fed into a S/P converter to create N parallel symbol streams. We label the individual symbol out of the converter and transmitted on the kth subcarrier in the mth subsymbol $d_{k,m}$. Each $d_{k,m}$ is upsampled by a factor N, i.e., $N-1$ zeros are appended. These upsampled sequences are filtered with $g_{k,m}(n)$, the pulse shaping transmit filter corresponding to the data symbol $d_{k,m}$, where

$$g_{k,m}(n) = g\left[(n-mK)\bmod N\right]e\left(-j2\pi\frac{k}{K}n\right) \quad (7.16)$$

and where $g_{0,0}(n)$ is the pulse shaping prototype filter. Each $g_{k,m}(n)$ is a time and frequency shifted version of the prototype filter $g_{0,0}(n)$. The outputs are thus at the designated subcarrier frequency. The baseband transmit signal in the digital domain $x(n)$ is obtained by the summation of all transmit signals and is thus given by

$$x(n) = \sum_{m=0}^{M-1}\sum_{k-0}^{K-1} d_{k,m}g_{k,m}(n) \quad (7.17)$$

Typically, the prototype filter is a square root-raised cosine-based filter. The resulting signals of all subcarriers are summed to create the GFDM signal. The output of the summer is fed out serially and a CP is added. Since only one CP is used for M time slots, spectral efficiency is improved further compared to CP-OFDM. We note that because the modulation symbols are now filtered, orthogonality in the OFDM sense is lost, and care must be taken to ensure that matched

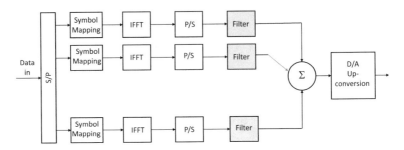

Fig. 7.30 UFMC transmitter block diagram

filtering per symbol is achieved at the receiver if the signal-to-noise ratio per subcarrier is to be maximized. GFDM achieves better OOB leakage suppression compared to CP-OFDM. However, it requires a more complicated receiver to address ISI and ICI, and the prototype filter may require more complex modulation, OQAM, for example.

7.5.3 Universal-Filtered Multi-carrier (UFMC)

With *universal-filtered multi-carrier* (UFMC) [18], subcarriers are grouped into sub-bands that are then filtered, unlike FBMC where individual subcarriers are filtered. The filtering is such as to reduce out of band emissions. The number of subcarriers and hence filter parameters is typically common per sub-band. Here CP can be eliminated, and the additional symbol duration that it would normally employ used to accommodate the sub-band filters. At the transmitter, first the complex modulation symbols are divided into groups and each group processed by an IFFT processor. Each IFFT output is then filtered. The outputs of the sub-band filters are then summed and the composite signal treated as a single OFDM transmission, i.e., it goes the D/A and upconversion processes. At the receiver, the composite signal is downconverter to baseband and processed as a standard OFDM baseband signal. Shown in Fig. 7.30 is the block diagram of an UFMC transmitter. Though UFMC offers good spectral efficiency relative to CP-OFDM, its implementation is considerably more complex in both transmitter and receiver.

7.6 Non-orthogonal Multiple Access (NOMA)

OFDMA and DFTS-OFDM are orthogonal multiple-access techniques where information for each user is assigned an exclusive subset of available frequency and time, thus resulting in no mutual interference among users and relatively simple transceiver design. These techniques have served 4G requirements for capacity well. However, compared to 4G, 5G has a much higher need for high volume services such as video streaming, cloud-based services, etc. One way to aid in meeting this increased capacity need would be to implement *non-orthogonal multiple access* (NOMA). NOMA achieves, at least in theory, higher spectrum efficiency and hence higher capacity by allocating, within the same cell, the same frequencies at the same time to multiple users. Many NOMA techniques have been proposed for application in 5G networks, and these are well summarized in [19]. To get a sense of how NOMA can influence total capacity, one such technique is addressed below, namely, power domain NOMA.

Power domain NOMA employs *superposition coding* (SC) at the transmitter and *successive interference cancellation* (SIC) at the receiver. Users operate with the same frequency resource at the same time and are superimposed in the power domain, differentiated from each other by their transmitted power level. At a given receiver, the received signal is the composite of all the signals transmitted. The first signal that SIC decodes is the strongest one, while all others are treated as interference. Here, by decoding we mean separating the strongest signal from the rest. This first decoded signal is then subtracted from the composite received signal, and if decoding is perfect, then the remainder represents all other signals. The process is repeated until it decodes its desired signal.

In DL NOMA the allocation of power at the BS is critical to the creation of sufficient differences in the power of the signal created for each UE so that signal separation at each receiver via SIC is effective. Normally, the most power is allocated to the UE located farthest from the BS and the least power to the UE closest to the BS. All UEs receive the same composite signal. The UE farthest away decodes the signal intended for it first since this signal is the strongest component of the composite signal. The UE nearest the BS, on the other hand, must first successively decode all signals other than that intended for it before finally decoding its own intended signal. To get a sense as to how power domain NOMA works in the downlink, consider the simple three user system depicted in Fig. 7.31, where the signal dedicated to each UE is OFDM modulated and occupies the entire transmission bandwidth. Here we see that UE3, the UE farthest away from the BS, simply demodulates S_{U3}, its intended signal, as S_{U3} is the strongest received signal. On the other hand, UE1, the UE closet to the BS, must a) first decode S_{U3} via SIC and then

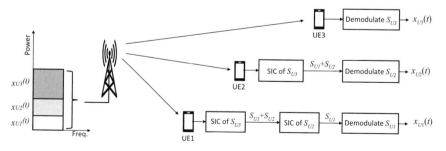

Fig. 7.31 Downlink power domain NOMA

subtract it from the composite to be left with $S_{U1} + S_{U2}$, the signals intended for UE1 and UE2 respectively, b) decode S_{U2} via SIC and subtract it from $S_{U1} + S_{U2}$ to be left with S_{U1}, and c) demodulate S_{U1}.

In the general case, if:

N is the number of mobile units
$b_n(t)$ is the individual OFDM transmitted waveform destined to the mobile unit UE$_n$
P_T is the total power available at the transmitter
f_n is the power allocation coefficient, i.e., the fraction of total power allocated to UE$_n$ $\sum_{1}^{N} f_n = 1$

$$P_n = f_n P_T$$

Then the composite signal transmitted by the BS is given by

$$b(t) = \sum_{n=1}^{N} \sqrt{P_n} b_n(t) \tag{7.18}$$

Then, if:

h_n is the complex valued channel coefficient between the nth UE and the BS
B is the overall transmission bandwidth
N_0 is the power spectral density of the additive white Gaussian noise at each UE

the received signal at UE$_n$ is given by

$$r_n(t) = b(t)h_n + N_0 B \tag{7.19}$$

For the UE located furthest from the BS, UE$_n$, being allocated the most power, it simply demodulates its own signal, treating all the other signals as noise. Its signal-to-interference and noise ratio is therefore given by

$$SINR_N = \frac{P_N |h_n|^2}{N_0 B + \sum_{i=1}^{N-1} P_i |h_n|^2} \tag{7.20}$$

For the UE located closest to the BS, UE$_1$, it will, using SIC, eliminate all the other signals before it can demodulate its own. Thus, assuming perfect SIC, the signal to be demodulated will only be corrupted by Gaussian noise, and hence its SINR is given by

$$SINR_1 = \frac{P_1 |h_1|^2}{N_0 B} \tag{7.21}$$

For the general case, for UE_n, $n \neq 1$, the SINR is given by

$$SINR_n = \frac{P_n |h_n|^2}{N_0 B + \sum_{n=1}^{n-1} P_n |h_n|^2} \tag{7.22}$$

Thus, by the Shannon Hartley theorem (Eq. 6.4), the throughput (bps) for data to each UE, where there is both interference and noise, is given by

$$R_n = B \log_2 \left(1 + SINR_n\right) \tag{7.23}$$

In OFDMA, there is no interference between signals as they are orthogonal, i.e., utilizing different frequencies. Here, where the total bandwidth and power is shared equally among the UEs so that:

The transmission bandwidth allocated to each UE is given by $B_n = B/N$
The power allocated to each UE is given by $P_n = P_T/N$

then the signal-to-noise ratio SNR for each UE is given by

$$SNR_n = \frac{P_n |h_n|^2}{N_0 B_n} \tag{7.24}$$

and the throughput given by

$$R_n = B_n \log_2 \left(1 + SNR_n\right) \tag{7.25}$$

The total capacity for both NOMA and OFDMA is given by

$$R_T = \sum_{n=1}^{N} R_n \tag{7.26}$$

In UL NOMA there is only one receiver at the BS, and this receiver uses, as in the DL, SIC to decode the incoming signals. The power transmitted by each UE is limited only by the maximum specified, and all users power can be independently adjusted by the BS. Assuming UEs are well distributed in the cell coverage area, the received power levels from the different UEs should be well separated with UEs all operating at nominal power. However, given the necessity to maintain distinctive power levels of all the signals arriving at the receiver in order to ensure effective SIC, the ability of the BS to adjust the individual UE's output power is desirable. This ability is all the more desirable given that individual UEs are likely to be in motion, thus constantly changing the level of their transmitted signal received at the BS. The strongest received signal power will be the first to be decoded, and under normal circumstances this signal is likely to emanate from one of the UEs closest to the BS. On the other hand, the weakest received signal will be the last to be decoded and under normal circumstances will likely emanate from one of the UEs farthest from the BS.

In situations where the BS controls the UEs output power level, if P_m is the UEs maximum transmitter power, assumed to be the same for all, then the power transmitted from each UE is given by $P_n = f_n P_m$, where f_n is the power allocation coefficient set by the BS, and the composite received signal at the BS is given by

$$r = \sum_{n=1}^{N} h_n \sqrt{P_n} b_n (t) + N_0 B \tag{7.27}$$

where here $b_n(t)$ is the individual OFDM waveform transmitted from UE_n and N_0 is the PSD of the Gaussian noise at the BS receiver input.

The BS decodes the signals from the UEs in a sequential fashion according to their power coefficients. For the decoded signal for UE_n, $n \neq 1$, the SNR is given by

$$SNR_n = \frac{P_n (h_n)^2}{N_0 B + \sum_{i=1}^{n-1} P_i (h_i)^2} \tag{7.28}$$

And for $n = 1$

$$SNR_1 = \frac{P_1 (h_1)^2}{N_0 B} \tag{7.29}$$

Equation 7.28 looks remarkably similar to Eq. 7.22 for the DL, but it is not in two important way. First, P_n is not constrained in the same way as in the DL, limited only by the maximum possible. Second, the signals being treated as interference each emanate from a different location and thus suffer a different channel coefficient. As with the DL, the throughput from each UE to the BS is given by Eq. 7.25, and the total capacity is given by Eq. 7.26.

To gain an intuitive understanding of how NOMA can outperform OFDMA regarding capacity, we consider two simple theoretical downlink cases, one with NOMA, the other with OFDMA, both with two users, and with user locations and hence channel coefficients the same for both cases.

For the NOMA version:

- Assume $P_1 = 1/5$, $P_2 = 4/5$, $h_1 = 1$, $h_2 = 1/2$, $SNIR_1 = 10 = 10$ dB, and bandwidth per UE equal total bandwidth B. With these assumptions the SNR (interference excluded) at each UE is the same as the higher power directed at UE_2 counteracts the added channel loss.
- Then, by Eq. 7.21, we find that $SNIR_1 = 1/(5N_0 B)$, which we have assumed equals 10. Hence $N_0 B = 1/50$.
- And, by Eq. 7.22, we find that $SNIR_2 = 1/(5N_0 B + 0.25) = 20/7 = 2.86$.
- Hence, by Eqs. 7.23 and 7.26, it can be shown that total capacity $R_{t(NOMA)} = 3.75B$.

For the OFDMA version:

- Assume $P_1 = 1/2$, $P_2 = 1/2$, $h_1 = 1$, $h_2 = 1/2$, $SNIR_1 = 10 = 10$ dB, and bandwidth per UE = $B/2$.
- Then by Eq. 7.24, we find that $SNIR_1 = 1/(N_0B) = 50$ and $SNIR_2 = 1/(4N_0B) = 7.5$.
- Hence, by Eqs. 7.25 and 7.26, it can be shown that $R_{t(OFDMA)} = 3.26B$.
- Thus $R_{t(NOMA)}$ is greater than $R_{t(OFDMA)}$ by 15%.

Clearly this simple example, though showing better NOMA capacity relative to OFDMA, does not prove the case for NOMA, but, at the very least, it indicates its potential. In fact, it has been shown in [20, 21] and in a number of other rigorous analyses that in general, under the right conditions, the throughput for DL and UL power domain NOMA using OFDM is greater than or equal to that for OFDMA, hence interest in its use.

An alternative to NOMA is hybrid NOMA, where the users in the network are divided into groups and the groups are allocated different frequency resource blocks. For a NOMA network with a large number of users, SIC can become highly repetitive and time-consuming for the users having to decode most all other signals before decoding their own. Hybrid NOMA minimizes this problem but at the trade-off of lower throughput.

Many NOMA realization techniques were considered as a study-item for 5G NR. However, it was decided not to continue with it as a 5G work-item but rather table it for further study and possible application in the next generation of mobile networks [22]. This decision was driven largely by the implementation complexity of current proposed schemes in return for negligible performance improvement over MU-MIMO (Sect. 8.3.4). The hope is that in time, with further study, complexity can be minimized and realizable performance relative to MU-MIMO become compelling.

7.7 Summary

In this chapter we explored multi-carrier-based multiple-access techniques as applicable to 5G NR. First, we reviewed OFDM, a multi-carrier technique whose basic structure lends itself to form the basis of multiple-access techniques. Two such techniques are employed by 5G NR, namely, OFDMA (sometimes referred to as CP-OFDM) and DFTS-OFDM, and these techniques were studied. Next, three other multiaccess schemes that were considered for application in 5G NR but ultimately not accepted, namely, FBMC, GFDM, and UFMC, were introduced. Finally, we took a high-level view of NOMA as it envisaged for possible future application in 6G or beyond. The motivation to explore the use of NOMA is that, by transmitting signals that are non-orthogonal to each other, it holds the promise of greater network capacity, i.e., greater data throughput capability.

References

1. Sari H et al (1995) Transmission techniques for digital terrestrial TV broadcasting. IEEE Commun Mag 33(2):100–109
2. Tosato F, Bisaglia P (2002) Simplified soft-output Demapper for binary interleaved COFDM with application to HIPERLAN/2. In: IEEE international conference on communications, conference proceedings, New York
3. Sesia S et al (2011) LTE-the UMTS long term evolution: from theory to practice, 2nd edn. John Wiley and Sons, Ltd., West Essex
4. Bisht M, Joshi A (2015) Various techniques to reduce PAPR in OFDM systems: a survey. Int J Signal Process Image Process Pattern Recognit 8(11):195–206
5. Heiskala J, Terry J (2001) OFDM wireless LANs: a theoretical and practical guide. SAMS Publishing Indianapolis, Indiana
6. Omri A (2019) Synchronization procedure in 5G NR systems. IEEE Access, Digital Object Identifier. https://doi.org/10.1109/ACCESS2019.2907970
7. Leeson DB (1966) A simple model of feedback oscillator noise spectrum. Proc IEEE 54(2):329–330
8. Zaidi A (2018) 5G physical layer: principles, models and technology components. Academic Press, London
9. Demir A et al (2002) Phase noise in oscillators: a unifying theory and numerical methods for characterization. IEEE Trans Circuits Syst I Fund Theory Appl 47(5):655–674
10. Schenk T et al (2005) Multiple carriers in wireless communications. Tijdschrift Ned Elektron Radiogenoot 70(4):112–123
11. Qi Y et al (2018) On the phase tracking reference signal (PT-RS) design for 5G new radio (NR). In: 2018 IEEE vehicular technology conference (VTC-Fall)
12. Singh S, Mishra S (2014) Analysis of PAPR on DFT-OFDMA systems. Int J Comput Appl 95(5):25–28
13. Morais DH, Inventor (2013) Quadrature amplitude modulation via modified-square signal point constellation. United States Patent, Patent No. US 8,422,579 B1, 16 April 2013
14. Morais D H, Inventor (2014) Hard and soft bit demapping for QAM non-square constellations. United States Patent, Patent No. US 8,718,205 B1, 6 May 2014
15. Bellanger M, Project Coordinator (2010) FBMC physical layer: a primer. PHYDYAS, June 2010
16. mmMAGIC (2016) Deliverable D4.1, Preliminary radio interface concepts for mm-wave mobile communications. Ver. 3.0, June 2016
17. Michailow N et al (2014) Generalized frequency division multiplexing for 5th generation cellular networks. IEEE Trans Commun 62(9):3045–3061
18. Vakilian V et al (2003) Universal-filtered multi-carrier technique for wireless systems beyond LTE. In: 2013 IEEE Globecom 2013 workshop – broadband wireless Access, pp 223–228
19. Wu Z et al (2018) Comprehensive study and comparison on 5G NOMA schemes. IEEE Access 6:18511–18518
20. Aldababsa M et al (2018) A tutorial on nonorthogonal multiple access for 5G and beyond. Hindawi Wirel Commun Mobile Comput 2018:Article ID 9713450
21. Tse D, Viswanath P (2005) Fundamentals of wireless communication. Cambridge University Press, Cambridge, UK
22. Makki B et al (2020) A survey of NOMA: current status and open research challenges. IEEE Open J Commun Soc 1:179–189

Chapter 8
Multiple Antenna Techniques

8.1 Introduction

Multiple antenna transmission is truly one of the key pillars supporting 5G aims of increased capacity and coverage. Such techniques have been well developed in 4G LTE systems. 5G NR builds on the progress made in 4G enabling reliable transmission over a much wider range of frequencies. In this chapter we will at a high level see how these techniques are applied in 5G NR. First, however, we will review the basics of such techniques.

Common to all multiple antenna techniques is the use of multiple antennas at the transmitter, at the receiver, or both, together with intelligent signal processing and coding. These techniques can be broken down into the following three main categories:

- Spatial diversity (SD) multiple antenna techniques: Diversity provides protection against deep fading by combing signals which are unlikely to suffer deep fades simultaneously.
- Spatial multiplexing multiple-input, multiple-output (SM-MIMO) techniques: SM-MIMO permits, in general, the transmission of multiple data streams using the same time/frequency resource thus improving spectral efficiency.
- Beamforming multiple antenna techniques: At the transmit end, beamforming permits the focusing of transmitted power in a given direction and thus increasing antenna gain in that direction, and at the receive it permits the focusing of the antenna directivity and hence gain in a given direction.

8.2 Spatial Diversity Multiple Antenna Techniques

Spatial diversity (SD) is enacted at receive end (*receive diversity*) by combining signals from multiple receive antennas and enacted at transmit end (*transmit diversity*) by transmitting signals via multiple antennas. When one transmitter antenna

© Springer Nature Switzerland AG 2020
D. H. Morais, *Key 5G Physical Layer Technologies*,
https://doi.org/10.1007/978-3-030-51441-9_8

feeds multiple receiver antennas, the system is referred to as a *single-input, multiple-output* (SIMO) one. When multiple transmitter antennas feed one receiver antenna, the system is referred to as *multiple-input, single-output* (MISO) one. Finally, when multiple antennas are employed at both transmit and receive ends to transmit *the same information*, the system is referred to as a *SD multiple input, multiple output* (SD-MIMO) one. Figure 8.1 shows block diagrams of the various configurations of spatial diversity.

The basic principle behind spatial diversity is that physically separated antennas receive signals or transmit signals that travel over different paths and are thus largely uncorrelated regarding fading. As a result, these signals are unlikely to fade simultaneously. Thus, by carefully combining them, average *signal-to-noise ratio* (SNR) or average *signal to interference and noise ratio* (SINR) of an SD system is improved relative to a SISO one. In the case of base-station antennas in a typical macro cellular environment, antenna separation on the order of ten wavelengths is typically needed to ensure low mutual fading correlation. For a mobile terminal in a similar environment, however, antenna separation on the order of only half a wavelength is often sufficient to achieve an acceptably low mutual fading correlation, given a large separation at the base station. SD can be used to achieve, with no increase in bandwidth, either of the following:

- Increased reliability via a decrease in average bit or packet error rate
- Increased coverage area
- Decreased transmit power

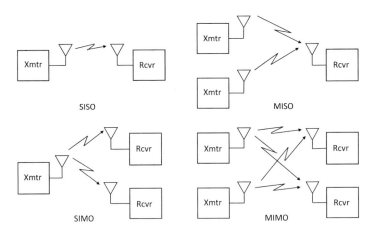

Fig. 8.1 Spatial diversity configurations

8.2.1 SIMO Maximum Ratio Combining Scheme

Many techniques can be used to combine signals received on multiple antennas from a single transmitting antenna thus creating a SIMO system. However, the most popular one is maximum ratio combining (MRC) and has been around for many decades. Here the received signals are combined so as to minimize the combined signal's fading effects by maximizing its SNR. Figure 8.2 shows an MRC system combining two fading signals so as to produce one with reduced fading.

8.2.2 Spatial Diversity MISO

MISO diversity is a more recent approach and is attractive in PMP systems as with its use diversity antennas only have to be installed at the BS. MISO diversity schemes are often characterized as either *closed loop* or *open loop*. Such closed loop systems require knowledge of the transmission channel at the transmitter. By knowledge we mean awareness of the channel's amplitude and phase characteristics over the channel's bandwidth. This is often referred to as *channel state information* (CSI). Such open loop systems, on the other hand, do not require this knowledge. Many transmit diversity schemes have been proposed. In a highly mobile environment, the channel changes quickly. Thus, closed loop MISO schemes tend to be applied primarily in fixed and low mobility scenarios. Open loop MISO systems, however, by not requiring up-to-date channel knowledge, operate well in a highly mobile environment.

8.2.2.1 Cyclic Delay Diversity

Cyclic delay diversity (CDD) is an open loop transmit diversity scheme. By the addition of delay to the additional transmit path or paths, it purposefully creates at the receiver a combined signal with additions and cancellations across the transmission frequency spectrum resulting in peaks and nulls. With OFDM, CDD involves sending the same set of OFDM symbols on the same set of subcarriers from multiple antennas, but with a different delay on each antenna. It operates block wise and

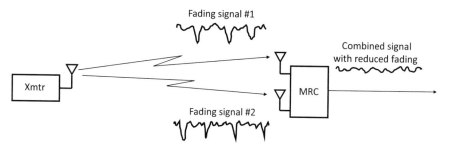

Fig. 8.2 MRC system combining fading signals

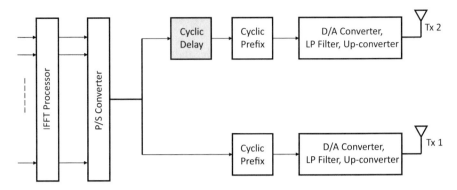

Fig. 8.3 CDD transmitter block diagram

applies cyclic shifts, rather than linear delays, to the different antennas. This creates the desired delay effect because a cyclic shift of the time domain signal is equivalent to a phase shift in the frequency domain. To guarantee that the delay is cyclic, it is applied before the cyclic prefix is added. Figure 8.3 shows a block diagram of a CDD transmitter.

In OFDM, cyclic delay is applied in the time domain by removing last n IDFT samples of a symbol and applying them to the start of symbol. Figure 8.4 illustrates the symbol structure of a three antenna CDD system showing cyclic delay introduced to the symbols destined for antenna 2 and 3 relative to the symbol destined to antenna 1. Because cyclic delay is added before the CP, no time delay is introduced prior to transmission, and thus there are no limits to the cyclic shift. Further, and importantly, there is no additional complexity in the receiver as the cyclic shifted signals appears as a multipath one which the receiver has already been designed to handle. CDD helps ensure that destructive fading is limited to individual subcarriers rather than an entire transport block. Thus, BER is not constant over the subcarriers. For uncoded transmission with CDD, the average BER will be approximately the same as in a flat faded channel. When the transmission block is encoded, however, with, for example, LDPC coding, then the coding process responds to the available frequency diversity, and the BER performance is improved.

8.2.2.2 Space-Time Block Coding

A very effective and one of the most popular forms of open loop MISO systems are ones employing *space-time block coding* (STBC) or variations thereof. With STBC, invented by Alamouti [1], the same data is sent via multiple antennas, but each stream coded differently. At the receiver, STBC algorithms and channel estimation techniques are then used to achieve both diversity and coding gain. In this scheme, the basic idea is to maximize use of both space and time diversity. Information is sent on two or more transmit antennas, with each antenna being fed by its own transmitter. With two transmit antennas, both antennas transmit in the same time slot two different signals, S_1 and S_2 say. They then transmit $-S_2^*$ and S_1^* in the next

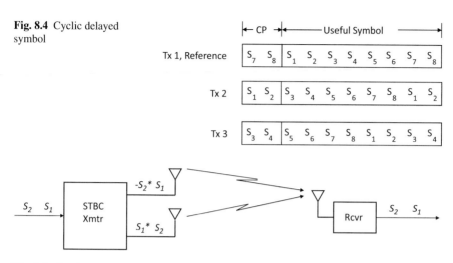

Fig. 8.4 Cyclic delayed symbol

Fig. 8.5 STBC SD-MISO system

time slot, where S_x^* is complex conjugate of S_x. Figure 8.5 shows such a system. The net effect is that the transmission rate is the same as if only one transmitter was being used. The reason for swapping signals from one time slot to the next is to statistically diversify the effect of the channel on information and thus increase the chance of correct signal reconstruction. The receiver waits for the received signals contained in the two consecutive time slots and then combines these signals via a few relatively simple computational operations to create estimates of the original signals. Note, however, that to accomplish this estimate of the original signals requires knowledge of the individual channels that the signals traversed. These estimates are sent to a "maximum likelihood detector" which outputs final estimate of the two signals. This scheme achieves the same diversity advantage as a single transmitter-two receiver one employing maximum ratio combining, reducing required fade margin by 3 dB under stable conditions, but by more under rapidly fading conditions. In OFDM/OFDMA systems, channel estimation is normally achieved via preambles or pilot tones which are split between the transmit antennas.

8.2.2.3 Space-Frequency Block Coding

Space-frequency block coding (SFBC), another open loop transmit diversity scheme, and inspired by the fundamentals of STBC, requires OFDM-type transmission. In this scheme, the basic idea is to maximize the use of both space and frequency diversity (as opposed to space and time diversity with STBC), the latter by transmitting the same data on two different subcarriers and hence on different frequencies. Transmission of the same data on different frequencies statistically diversifies the effect of the channel on the data and thus increases the chance of correct data reconstruction. Like the STBC method, this scheme transmits two complex symbols, S_1 and S_2, each over two transmit antennas, with each antenna being fed by its own transmitter. SFBC transmits data on allocated subcarriers from both antennas in the format shown in the Fig. 8.6. Like STBC, net effect is that transmission rate is same as if only one transmitter was being used.

8.2.3   Spatial Diversity MIMO

The use of multiple receive antennas along with STBC or SFBC multiple transmit antennas creates a spatial diversity MIMO (SD-MIMO) system. Such systems increase the number of diversity paths and hence diversity gain which further improves reception of the transmitted signals. For a system with N_T transmit antennas and N_R receive antennas, the number of diversity paths is now $N_T \times N_R$. In such systems the receiver processing is via an MRC combiner followed by a maximum likelihood detector. SD-MIMO has found application in 4G LTE systems.

8.3   Spatial Multiplexing MIMO

MIMO systems are very interesting because of their dual capability. They can be used, as indicated above, to provide very robust spatial diversity. They can also, however, be used to increase capacity via *spatial multiplexing* (SM), with no additional transmit power or channel bandwidth compared to a SISO system. Such a scheme is referred to as *spatial multiplexing multiple-input, multiple-output* (SM-MIMO). It is important, therefore, to distinguish between SD-MIMO and SM-MIMO. To simplify presentation going forward, "spatial diversity MIMO," when abbreviated, will be indicated as SM-MIMO, while "spatial multiplexing MIMO," when abbreviated, will be indicated as simply MIMO.

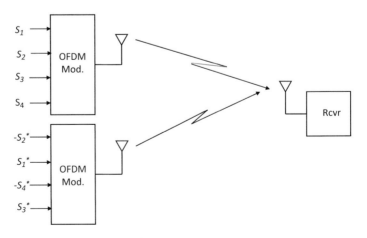

Fig. 8.6 SFBC SD-MISO system

8.3.1 MIMO Basic Principles

Consider a MIMO system with N transmit antennas communicating with M receive antennas, where $M \geq N$. At the transmitter, input data is divided via a S/P converter into n substreams. These substreams are then encoded and used to modulate N carriers, each occupying the same channel, and these modulated carriers feed the n antennas. Thus, a "matrix" channel consisting of $N \times M$ spatial dimensions exists within the same assigned bandwidth. Clearly, MIMO increases throughput n times compared to SISO, SIMO, MISO, or SD-MIMO ones. Successful MIMO requires transmission so that the received signals are highly decorrelated. This can be achieved by transmitting the different signals on different polarizations and/or transmitting them over a propagation channel that is rich in multipaths that result from signals bouncing and scattering off nearby objects such as buildings, trees, and cars. At receive end, each of the M antennas picks up all the transmitted substreams and their many images, all superimposed on each other. However, because each substream is launched from a different point in space, each one is scattered slightly differently than the rest. These differences in scattering are key to successful spatial multiplexing transmission. They decorrelate paths taken by the separately transmitted substreams. This minimizes destructive combining at receiver, allowing individual substreams to be identified and recovered via sophisticated signal processing. Recovered substreams are decoded and recombined to recreate original signal. Figure 8.7 shows a capacity-enhancing one-way 2×3 MIMO system.

An analysis of the algorithm used to demultiplex a 3×3 MIMO system can give insight as to how it works. Consider the 3×3 MIMO channel shown in Fig. 8.8. Ideally this system results in nine independent transmission channels, each with a scalar coefficient and hence flat fading conditions. Let the coefficient between the receive antenna A_{Ry} and the transmit antenna A_{Tx} and be H_{yx}. The composite signal at each receiving antenna, A_{Ry}, is the addition of the signals received from the three transmit antennas. Thus, for example, $R_1 = T_1 H_{11} + T_2 H_{12} + T_3 H_{13}$. At the receiver, the nine coefficients can be identified through training symbols, achieved, for example, by transmitting a training sequence from one antenna at a time, while the other

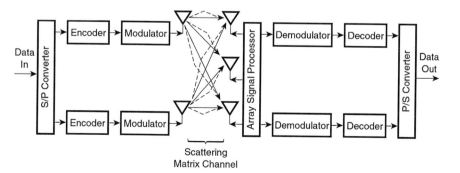

Fig. 8.7 A 2×3 MIMO system

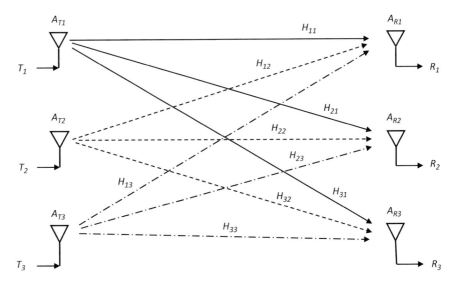

Fig. 8.8 A 3 × 3 SM-MIMO channel

two transmit antennas are idle. Each transmission allows the determination of three of the coefficients; hence the three independent transmissions allow determination of all nine coefficients. As a result, we have three equations for the three received composite signals and three unknowns, T_1, T_2, and T_3. We are thus able to compute the unknowns.

More generally, for a MIMO system with N transmit antennas and M receive antennas, we can describe demultiplexing via matrix algebra as follows:

$$y = Hx + n \tag{8.1}$$

where y is the $M \times 1$ matrix of the resulting signals at the M receive antennas, x is the $N \times 1$ matrix of the signals sent from the N transmit antennas, n is the $N \times 1$ matrix of the noise at the N receive antennas, and H is the $M \times N$ channel matrix of the form:

$$H = \begin{pmatrix} h_{11} & \cdots & h_{1N} \\ \vdots & \ddots & \vdots \\ h_{M1} & \cdots & h_{MN} \end{pmatrix} \tag{8.2}$$

If we ignore the noise, then:

$$\hat{x} = \hat{H}^{-1} y \tag{8.3}$$

where $\hat{x}$ is the estimate of the transmitted signals and $\hat{H}^{-1}$ is the transpose of $\hat{H}$, the receiver's estimate of the channel matrix. Clearly, with no noise and perfect channel matrix computation, we get perfect signal recovery. However, if the SINR is low and or the decoded channel matrix values vary much from reality, then signal recovery will be poor.

We note that in OFDM systems, decoding is done on a subcarrier by subcarrier basis; hence the assumption in our example of scalar coefficients in the channel matrix is essentially correct, as individual subcarriers are normally sufficiently narrow as to be subjected to flat fading. Since successful transmission is dependent on multipath scattering, not all UEs will necessarily achieve capacity enhancement. Assuming strong scattering, those close to BS are likely to be able to achieve it, given high probability here of good SINR. However, those far from BS or those with near line of sight with BS where scattering is weak are unlikely to be able to benefit from capacity enhancement. For MIMO systems that offer capacity enhancement, one way of providing back up in event of unsuitable channel conditions is to adaptively switch to STBC or SFBC (SD-MIMO) whenever such conditions arise. With this switch, capacity is decreased in exchange for reliability. Such switching referred to as adaptive MIMO switching (AMS) and leads to optimized spectral efficiency at both low and high SNR regions.

8.3.2 Antenna Array Adaptive Beam Shaping

Adaptive beam shaping is a technique whereby an array of antennas is used adaptively for reception, transmission, or both in a way that seeks to optimize the transmission over the channel. What we refer to here as beam shaping is also loosely referred to as beamforming, but as beamforming has recently taken on a more specific connotation, the term beam shaping will be used here. In a PMP environment, antenna arrays can be located at both the base station and remote units. In the case of base station arrays, unlike omnidirectional or sectorized antennas that cover an entire cell or sector respectively with almost equal energy, regardless of the location of the remotes, adaptive beam shaping arrays target an individual user equipment (UE) or multiple UEs. It is able to do the latter by creating multiple beams simultaneously, each beam directed to an individual UE. The shape of each beam can be dynamically controlled so that signal strength to and from a UE is maximized, by directing the main lobe in the direction of minimum path loss and side lobes in the direction of multipath components. Further, it can simultaneously be made to minimize interference by signals that arrive at a different direction from the desired by locating nulls in the direction of the interference. Thus, this technique can be made to maximize the SINR. Figure 8.9 shows the beams of an adaptive beam steering antenna array communicating with two remote stations in the presence of multipath and interference.

An *antenna array* consists of two or more individual antenna elements that are arranged in space and interconnected electrically via a feed network in such a fashion as to produce a directional radiation pattern. In adaptive beamforming, the phases and amplitudes of the signals in each branch of the feed network are

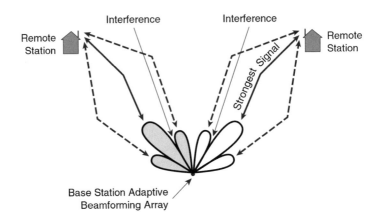

Fig. 8.9 Beams formed from an adaptive beam shaping antenna array communicating with two mobile units

adaptively combined to optimize the SINR. To demonstrate the basic principles of an antenna array, consider the simple M-element *linear equally spaced* (LES) array shown on Fig. 8.10a. This array is oriented on the x-axis with a spacing of Δx between adjacent elements and is shown receiving a plane wave from the direction (α, β). Figure 8.10b shows the elements and its feed network, and, to keep the analysis simple, a plane wave arriving in the x-y plane ($\beta = \pi/2$) is assumed. This results in the difference in length, Δl, between rays received by any two adjacent elements being

$$\Delta l = \Delta x \cos \alpha \tag{8.4}$$

and thus, the difference in phase, i.e., phase shift, $\Delta \phi_k$, between the ray received by the k th element and the first element (where $k = 0$), being

$$\Delta \phi_k = 2\pi k \frac{\Delta l}{\lambda} \tag{8.5a}$$

$$= 2\pi k \frac{\Delta x}{\lambda} \cos \alpha \tag{8.5b}$$

where λ is the wavelength of the received wave.

To maximize reception of this wave, the de-spreading weighting elements w_k in the feed network are adjusted such that the signals from all elements sum coherently in phase. In effect, all signals other than that received on element M-1 are delayed so as to have the same phase as that on element M-1. Thus, the signal on element 0, for example, is delayed by $\Delta \phi_M$. Note that since a wave arriving perpendicular to the x-axis would result in no phase shift between the received rays at the elements, then to maximize reception of such a wave requires no delay by any element. This example only shows phase being weighted, but in adaptive beamforming, as indicated

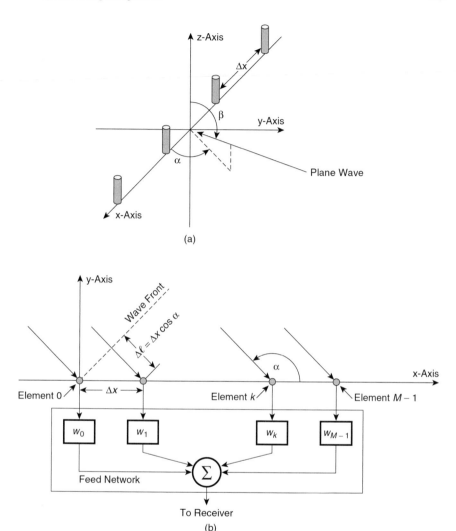

Fig. 8.10 M-element linear equally spaced array. (**a**) M-element array receiving a plane wave. (**b**) M-element array and feedback network ($\beta = \pi/2$)

above, both phase and amplitude are normally adjusted. In a PMP system, all BS antenna elements receive signals from all active remotes simultaneously. These signals are then processed by sophisticated DSP circuitry to create the element weights necessary to simultaneously generate a separate antenna pattern for each remote.

The example above gave some insight into how adaptive beamforming works as a receiving antenna system. With the antenna at the base station, this is good for upstream transmission, even if the signals arrive via NLOS routes. However, this is of little use if the downstream transmission is not similarly processed. To do this, information about the downstream paths is needed. With frequency division duplexing (FDD), this can only be made available via feedback paths from the remotes to the base station, which adds complexity and consumes some of the upstream bandwidth. For this reason, most systems that employ adaptive beamforming use time division duplexing (TDD). (FDD and TDD are described in Sect. 9.7.) This way, since the downstream and upstream paths are now identical, information necessary for effective downstream transmission is available from upstream transmission. For transmission, the adaptive array operates in the reverse of its reception mode. The signal to be transmitted is fed through a feed network that splits it into M components and applies a spreading weight to each, prior to feeding an antenna element, the spreading weight being the same as the de-spreading weight applied to that element. Thus, when these signals arrive at the intended remote, their weights are such that, even if arriving over a NLOS route, they combine with each other coherently.

Compared to standard omnidirectional or sector antennas at the BS and omnidirectional antennas at the UE, adaptive beamforming antennas, by their ability to focus beams, result in significantly increased coverage and capacity in both LOS and NLOS environments. We note that the gain, and hence the directivity, i.e., the degree to which the radiated energy is focused in a single direction, of a linear antenna array, is a direct function of the number of antenna elements. More specifically, for an array with N elements and half-wavelength element spacing, gain, G_{array}, is given by

$$G_{array} = 10\log_{10}(N) + G_e \tag{8.6}$$

where G_e is the gain of each element.

Thus, an antenna array with a large number of elements will be able to transmit a highly focused beam with high gain relative to a single element one. In array antennas, elements are typically half-wave dipoles and hence G_e equals 2.15 dB.

8.3.3 MIMO Precoding

By application of *precoding* at the transmit end, it is possible to improve MIMO system performance via mild beamforming. Here precoding refers to the linear combining of the original data streams (layers) in the transmitter. When properly applied, this results in an increase and/or equalization of the receiver SINR across the multiple receive antennas. In a precoded MIMO system, inverse operations are performed at receiver to recover the original un-precoded signals.

Figure 8.11 shows a simplified version of a 2 × 2 MIMO system with precoding. On the transmit side of this system, blocks of data bits enter the modulator which transforms these blocks to corresponding blocks of complex modulation symbols, for example, 16-QAM symbols or 64-QAM symbols. The modulator feeds the *layer mapper*. The layer mapper creates independent modulation symbol streams, each

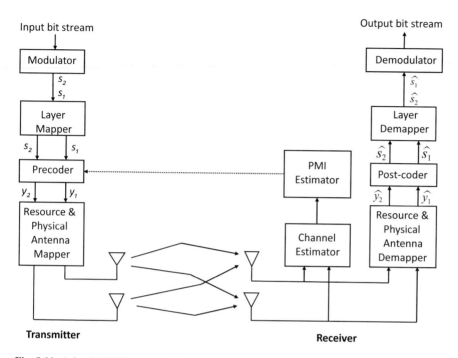

Fig. 8.11 A 2×2 MIMO system with a precoding

destined ultimately to an independent antenna. If the mapper is to create n streams, then every nth symbol is mapped to the nth layer. Thus, in our case under study, every second symbol goes to the second layer. The layer mapper feeds the precoder. In the precoder, the layers are combined adaptively via one of a set of weighting matrices. This set of matrices is referred to as a *codebook*. For a 2×2 configuration such as shown in Fig. 8.8, input layers S are multiplied by a weighting matrix W to generate the precoded signals Y given by

$$\begin{bmatrix} y_1 \\ y_2 \end{bmatrix} = \begin{bmatrix} w_{11} & w_{12} \\ w_{21} & w_{22} \end{bmatrix} \begin{bmatrix} s_1 \\ s_2 \end{bmatrix} \tag{8.7}$$

The output of the precoder goes to the resource and physical antenna mapper. Here the modulation symbols to be transmitted on each antenna port are mapped to the set of time/frequency resource elements allocated by the scheduler. By mapping we mean that the modulation symbols modulate specified subcarriers in specified time slots. Following resource mapping, the modulated subcarriers are mapped to specific antenna ports. On the receive side, the process is reversed. Note, however, that here estimates of the channel matrix H are made via information gleaned from received reference signals on the different antenna ports. These estimates are required to demultiplex the spatially transmitted signals. It is also used, importantly, to feed information back to the transmitter so that the optimum codebook matrix may be applied. From the *singular decomposition* (*SVD*) of *H*, it computes the optimal precoding matrix (SVD description beyond scope of this course). The receiver also knows the codebook. It thus determines which codebook matrix it deems most suitable under current path conditions. The receiver sends this matrix recommendation as a *precoder matrix indicator* (PMI) along with other channel data such as SINR to the transmitter where the final matrix decision is made. Because, in the system described, the receiver feeds back channel derived information to the transmitter, such a system is described as a *codebook based closed-loop precoding* one.

It is also possible to have *non-codebook-based transmission*. In the DL, if transmission is time division duplexed and hence the channel is reciprocal, the BS, rather than relying on a PMI sent by the UE, can generate transmit antenna weights from channel estimates obtained from UL transmissions. The same process can also be used in the UL to allow non-codebook-based transmission, but here it's the UE that generates its own transmitter antenna weights.

8.3.4 Single-User and Multi-user MIMO

A *single-user MIMO* (SU-MIMO) system is a spatial multiplexing one where, at any given time, transmission *on a given frequency resource* can only be between a BS and one UE. Note that this does not mean that transmission can only be to a single user at any given time, as transmission to other users can be realized via other available frequency resources. The goal of SU-MIMO is to allow the maximization of the individual user data rate. It can be implemented in both DL and UL utilizing the multiple layers specified. With single-user MIMO, the UE thus requires multiple antennas to take advantage of spatial multiplexing. Recall also that for spatial multiplexing between the BS and UE to be effective, the link needs to undergo strong multipath scattering.

A *multi-user MIMO* (MU-MIMO) system, in contrast to a SU-MIMO one, is one where, at any given time, transmission *on a given frequency resource* can be between a BS and multiple UEs. Thus, if need be, the full frequency resource can be used for transmission between the BS and each UE. The goal of MU-MIMO is the increase total cell throughput, i.e., capacity. It requires only a single antenna at the UE. The system is still a spatial multiplexing one, but here UE antennas are essentially distributed over several UEs. Now, in the DL, because each UE does not have to demultiplex several layers, the requirement for a channel rich in multipath is removed. In fact, line of sight propagation, which causes considerable performance problems with SU-MIMO, is not a problem here. MU-MIMO can be implemented in both DL and UL. In DL MU-MIMO the BS, via beam steering, transmits beams simultaneously to each current user. It can be open loop or closed loop. In a MU-MIMO system that processes signals to/from *n* UEs, *n* sets of *M* weighting elements are used at the BS to produce *n* outputs. Each set of *M* weights can null *M*-1 user signals besides maximizing transmission/reception of a particular desired signal. Thus, for successful multiplexing/demultiplexing of n signals, M must be $\geq$ n. MU-MIMO requires that UEs have sufficient angular separation in space as viewed by the BS so that the weighting network can differentiate between them. As a result, MU-MIMO is difficult to implement in mobile environments where UEs may be adequately separated one moment and be in the same direction from BS in the next. MU-MIMO provides a capacity increase as multiple users use the same frequency and time resources simultaneously. Note, however, that for efficient MU-MIMO, the interference between users must be kept low. This can be achieved by using beam shaping as discussed in Sect. 8.3.2 so that when a signal is sent to one user its nulls are formed in the direction of the other users. Figure 8.12 shows a MU-MIMO system where the number of UEs n equals 2 and the number of weighting elements per set, M, and hence number of BS antennas equal 3.

In DL MU-MIMO utilizing FDD a closed-loop system is required. The BS continually reviews the channel state information (CSI) sent up from all remotes and uses this information to determine weighting, but only schedules transmission to those remotes with favorable CSI, and to no more remotes than maximum number permitted. In DL MU-MIMO utilizing TDD, the UL and DL channels are reciprocal, and so no CSI from the remotes is required, and the system is thus open-loop. The BS determines the CSI of the UL channels, assumes this information is valid for the DL channels, and uses it to compute the necessary weights to result in constructive signal combination at each UE.

In UL MU-MIMO, also referred to as UL collaborative MIMO, UEs transmit collaboratively in the same time and frequency resource. If say two UEs are transmitting collaboratively, UL throughput is doubled compared to only a single UE transmitting. Note, however, that individual throughput is unchanged. For processing of the received signals, channel estimates are made from known reference

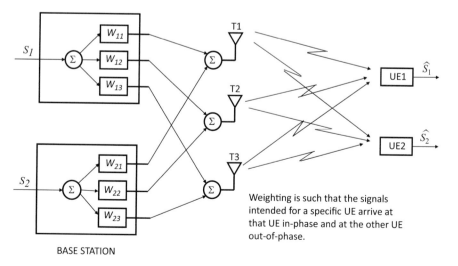

Weighting is such that the signals intended for a specific UE arrive at that UE in-phase and at the other UE out-of-phase.

BASE STATION

Fig. 8.12 A DL MU-MIMO system showing weighting elements

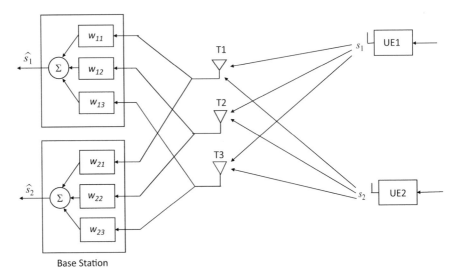

Fig. 8.13 An UL MU-MIMO system showing weighting elements

signals received. These estimates are then used to determine the weights necessary to cause constructive combination of the signals. Figure 8.13 shows a simplified UL MU-MIMO system.

8.3.5 Massive MIMO and Beamforming

The simplest description of *massive MIMO* (mMIMO) is that it is nothing more than MU-MIMO with a "massive" amount of BS antennas, certainly much more than employed in traditional MU-MIMO, and much more than there are users in the cell. How many BS antennas does it take to create a mMIMO system? There is no definition, but it's generally considered to be about 60 or more. The concept of mMIMO was first proposed in 2010 by Mazetta [2] with the assumption of TDD transmission. As with MU-MIMO, each UE can be allocated the entire bandwidth, and because of the large number of BS antennas, more independent data modulated signals can be sent out, and thus more UEs can be served simultaneously. Broadly speaking, we can think of mMIMO systems as ones with close to a hundred or more of BS antennas serving close to ten terminals or more.

In MU-MIMO systems such as used on LTE, the BS obtains DL channel knowledge by transmitting pilot signals to the EUs; the UEs use these signals to estimate the channel responses and then feed this information up to the BS. For two reasons, this approach is not implementable with mMIMO. First, to be strongly recognizable, the DL pilots must be mutually orthogonal between the various UE antennas. Thus, they must each be sent on an independent time/frequency resource. As a result, the DL pilot required time/frequency resource scales with the number of BS antennas and would thus be very large in a mMIMO system with hundreds of BS antennas. The second reason is that the number of channel responses that each UE estimates is also proportional to the number of BS antennas. Thus, the UL resources required to convey to the BS the channel responses would be very much larger than in conventional MU-MIMO. The solution to this problem is to use TDD so that the DL channels can be assumed to be the same as the UL ones and thus estimate the DL channel responses via UL pilots. Another advantage of UL pilot-based channel estimation is that all the excessive amount of computation and signal processing is done at the BS, not at the UEs.

As with MU-MIMO, the wavefronts emitted from each antenna are weighted so as to add constructively at the intended UE and destructively at other UEs. With mMIMO, because of the large number of antennas directed to the individual UE, coupled with the fact that there are much more BS antennas than likely active users, the overall signal become very directive. Another way of saying this is that the transmitted beam becomes very narrow. This creation of a narrow beam is what is referred to as *beamforming*. A multi-antenna array beamformed signal directed to a single antenna UE can be likened to a laser beam, whereas the more traditional MU-MIMO beam shaping signal can be likened to a flashlight beam. Beamforming is a good technique for situations where there is one dominating propagation path. This is the situation with LOS. It can also, however, be the case where there is no LOS but a strong specular reflection. In the latter situation, the transmitter can focus the beam at the point of reflection and the receiver focus its directivity at the point of reflection.

An important feature of mMIMO is *channel hardening*. When the number of BS antennas is large as with mMIMO, there is significant spatial diversity, and the channel hardens. With channel hardening [3], small-scale fading decreases leaving primarily only large-scale fading, thus the channel appears as a flat faded one only, with slow variation over time compared to small-scale fading and no rapid fading dips. With hardening the received signal power typically remains well above the thermal noise power at the receiver. As a result, the influence of small-scale fading and thermal noise can be largely neglected compared with inter cell interference.

Some benefits of mMIMO systems are:

- Cell capacity can be increased by at least an order of magnitude. This is because, with increased spatial resolution, more UEs can be served without an unacceptable build up in mutual interference between the UEs.
- Radiated energy efficiency can be improved significantly in both the DL and UL. This is because, with the laser like focus possible, little energy is wasted. In the DL, the effective BS antenna gain is so high that the radiated power per UE can be reduced by an order of magnitude or more. Likewise, in the UL, the effective BS antenna gain is so high that the individual UE transmitted power can be substantially reduced.
- Management of the multiple access layer is simplified as a result of channel hardening. With the received signal appearing flat across the frequency spectrum, frequency domain scheduling becomes unnecessary, and each UE can be allocated the entire bandwidth if required.

- System scheduling, power control, etc. can be carried out over the slower large-scale fading time scale, simplifying the management measurably.
- Because, again because of hardening, the UE is unlikely to be trapped in a deep fade, latency improves significantly as the UE doesn't have to wait for favorable propagation conditions for data to be sent, and requests fewer retransmissions of data due to corruption.
- Coverage is improved. Because of the more focused, higher radiated power to each UE, users experience a more uniform, higher data rate service throughout the cell coverage area.
- Features are highly applicable to millimeter wave bands. Propagation characteristics of these bands are generally poor. The high directivity and hence high antenna gain resulting from beamforming can help in overcoming this limitation.

8.3.6 Antenna Array Structure

In mMIMO systems, beamforming is accomplished via antenna arrays. These arrays enable high directivity beams and the ability to steer these beams over a range of angles in both the horizontal as well as vertical plane. In general, as was indicated in Sect. 8.3.2 above, the more antenna elements used, the higher the array gain. The beam is steered by controlling the weighting of smaller parts of the array. These smaller parts are called *subarrays*. A subarray typically has multiple antenna elements placed vertically, horizontally, or in a two-dimensional structure. The elements are normally half-wave dipoles, uniformly spaced typically half a wavelength apart, and may either be single polarized or dual polarized. Subarrays are placed side by side to create a uniform rectangular array. Figure 8.14 shows an array with 2 × 1 subarrays, each containing two dual-polarized elements. This array is steered by applying two dedicated radio chains per subarray, one per polarization. It has 16 subarrays and thus 32 transmit/receive RF points of interface.

Each subarray has a radiation pattern that describes its gain in different planes. The gain across the horizontal plane, for example, may be different from that across the vertical plane. The radiation pattern is a function of the structure of the subarray. The array gain in a given direction is thus the product of the subarray gain in that direction and the number of subarrays and is the gain achieved when all subarray signals in that direction are added constructively. The structure of the subarray determines the steerability of the total array in a given direction. Consider the array shown in Fig. 8.14. It has two 2 × 1 subarrays in the vertical direction per array column. It thus has two different weights (per polarization) than can be applied per column to influence directivity and gain in the vertical direction. Why not then always use a 1 × 1 subarray and thus have four different weights that can be applied in the vertical direction? This would give maximum directivity in the vertical direction. The answer is that the smaller the subarray, the more radio chains that must be connected to the total array and hence the higher the cost of implementation. The subarray structure is therefore typically chosen based on the topography to be

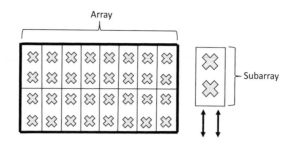

Fig. 8.14 ITU BS antenna array model

covered. If the spread in the vertical direction of users to be covered is high, as may be found in a dense urban environment, then more subarrays in the vertical direction is preferable. If, on the other hand, the spread of users in the vertical direction is low, as may be the case in suburban or rural areas, then less subarrays in the vertical direction is preferable. In the horizontal direction, the array shown in Fig. 8.14 has eight 2×1 subarrays per array row. It thus has eigth different weights (per polarization) that can be applied per row to influence directivity and gain in the horizontal direction. Directivity achievable in this direction is thus the maximum possible with this array structure.

8.3.7 Digital, Analog, and Hybrid Beamforming

One key aspect of BS multi-antenna beamforming implementation is where in the transmission chain weighting to effect beamforming should occur. On a high level three options can be defined, namely, digital beamforming, analog beamforming, and hybrid beamforming [4]. Following we will explore these options.

8.3.7.1 Digital Beamforming

Digital beamforming at the BS can be viewed simply as MU-MIMO described above, with precoding at the baseband level and antenna mapping to a large number of antennas. Multiple beams, one per user, can be simultaneously formed from the same set of antenna inputs. Figure 8.15 is a block diagram of the transmitting section of such a system. Here data streams destined to N_U users are inputted to the digital baseband processor and precoder. The precoder outputs N_T streams, one for each antenna. These streams then each go through DACs and are upconverted to RF, amplified, and fed to the antenna. We define the DAC and upconverter as an RF chain. Note that each "antenna" input here is not necessarily one antenna element. It is likely an input to a subarray, which, if it contains dual-polarized antennas, then has separate inputs per polarization and thus separate RF chains. Now that we have the big picture, let's take a closer look at what is happening. We start by assuming

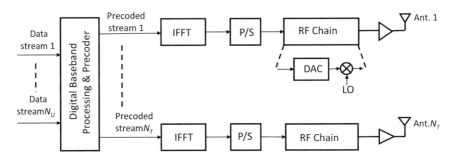

Fig. 8.15 Digital beamforming transmitter architecture

only one user and N_T antenna inputs. The user data creates modulation symbols, for example, QAM symbols, which feed the precoder. The precoder creates N_T modulation symbols streams, all modulated by the same data, but weighted differently such that the N_T signals transmitted by the antenna add coherently at the user terminal. The precoder outputs each go through an IFFT followed by a P/S process, thus creating N_T OFDM modulated signals. These signals then go through the DACs and are upconverted to the transmission frequency. Let's now assume that there are N_U users. The modulation symbols for each user are fed into the precoder which still produces N_T outputs. The difference here is that the OFDM signals now created are modulated by N_U users and are such that the signals at the antenna go off in N_U directions. As you can imagine, quite a complex process. For the handling of signals received at the BS array the structure is the same as shown in Fig. 8.15 but with signal flow and hence processes reversed. Thus, for example, DACs are replaced by ADCs and IFFTs are replaced by FFTs.

An RF chain for each antenna input provides very high performance and flexibility regarding spatial multiplexing, beam steering and directivity, interference suppression, etc. However, the price for this versatility is expense in terms of costly RF components and high-power consumption by both the RF components as well as the DACs, given that one RF chain is required per antenna input. DAC power consumption is a function of the sampling bandwidth which in turn is a function of the bandwidth being processed. At millimeter-wave frequencies, where RF costs are high and bandwidths likely large, these disadvantages become significant. Thus, a fully digitally formed beamforming system is unlikely in millimeter-wave bands but more likely in sub 6 GHz bands where RF costs and available bandwidths are lower.

8.3.7.2 Analog Beamforming

A block diagram of analog beamforming in the transmitting section of a BS is shown in Fig. 8.16. The key difference with digital beamforming is that the beamforming procedure takes place at RF on a single OFDM modulated signal versus taking place at baseband in a precoder on several modulated symbol streams. After

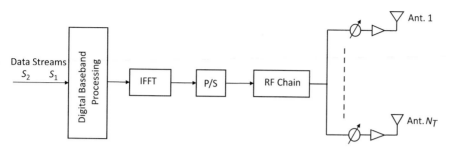

Fig. 8.16 Analog beamforming transmitter architecture

upconversion, the single OFDM signal is split into N_T streams that feed N_T antenna inputs via individual phase shift and gain adjustment elements. With these elements beam steering and side lobe suppression can be controlled. We note that this arrangement has only one RF chain, versus one per antenna input in the case of digital beamforming. Thus, beamforming is performed in the analog domain at the RF frequency. It is carried out on a per carrier basis. As a result, for DL transmission, it is not possible to simultaneously transmit multiple beams directed to different users located in different directions as can be done with digital beamforming. Instead, transmission to multiple users must be on a time division basis, leading to a lower overall throughput than with digital beamforming.

With analog beamforming the weights cover the entire bandwidth thus weighting cannot counter channel variations due to frequency selective fading. Analog beamforming is thus better applicable to situations where frequency selective fading is minimal, such as LOS links, those with a strong specular reflection, or those with a dominating cluster of multipaths with low angular spread so as to minimize relative path delay. Such situations are likely to be the case with millimeter wave networks employing small cells as opposed to sub 6 GHz networks where cells will be more likely to be large and suffer from multipath.

8.3.7.3 Hybrid Beamforming

Hybrid beamforming, as the name suggests, is a combination of analog and digital beamforming processes. It seeks to combine the advantages of and minimize the disadvantages of analog and digital beamforming architectures. There are two common architectures for realizing hybrid beamforming. One is referred to as fully connected and the other as partially connected.

Shown in Fig. 8.17 is the fully connected architecture. Here N_U user-directed data streams are processes and precoded to produce N_R RF chains where N_R is equal to or greater N_U but much less than N_T, the number of antenna elements. Each RF chain drives a set of N_T phase shift and gain controllers which in turn drive the N_T antenna elements. Thus N_R beams with full analog beamforming gain can be produced independently of each other, these beams being reinforced digitally via

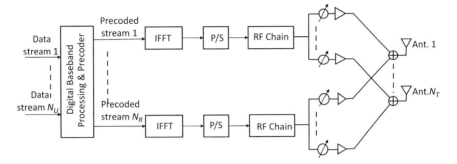

Fig. 8.17 Fully connected hybrid beamforming architecture

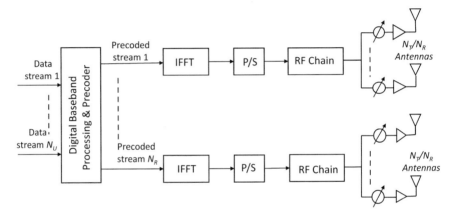

Fig. 8.18 Partially connected hybrid beamforming architecture

digital precoding. Note, however, that since the number of RF chains is now much lower than the number of antennas, there are less degrees of freedom for digital baseband processing than available with full digital beamforming.

Shown in Fig. 8.18 is the partially connected architecture. Here, as in the fully connected one, N_U user-directed data streams are processes and precoded to produce N_R RF chains where N_R is equal to or greater than N_U but much less than N_T. In this case, however, each RF output is connected to a partial section of the full array consisting of N_T/N_R antenna elements, with each element connected via a phase shift and gain controller. Analog beamforming can thus be produced at each partial section. However, the beamforming gain at each partial section is less than that achievable from the full array, as is possible with the fully connected architecture. If there was no precoding and N_R was equal to N_U, then this would simply be N_U analog beamforming transmitters but with each having less antenna elements than available from the full array. The difference here, however, is that the beams formed at the subarrays can be reinforced by applying digital precoding across partial sections of the full array. The partially connected structure has a lower hardware complexity compared to the fully connected one, but at the expense of reduced beamforming gain.

In both the fully and partially connected structures, the number of simultaneously supported streams is greater than with analog beamforming, where only one is supported, but less than with digital beamforming. However, hybrid beamforming is a more practical approach than full digital beamforming for millimeter wave cells because of the significant reduction in the number of RF chains and lower digital processing complexity.

8.3.8 Full-Dimension MIMO

Traditional beamforming schemes controlled the beam direction only in the horizontal plane, i.e., in two dimensions only. In Sect. 8.3.6 above, we alluded to being able to structure an antenna array to allow steerability of beams in both the vertical and the horizontal direction. When beams are so steered, then their direction can be controlled in three dimensions.

A MIMO system that allows beam steerability in three dimensions is referred to as a full-dimension MIMO (FD-MIMO) one [4]. This extra degree of freedom allows the more precise directing of a beam to a specific user resulting in higher average user throughput, less interference at a given UE from beams directed to other UEs, and improved coverage. These benefits of FD-MIMO require narrow targeted beam focusing and thus a large number of array elements. They further require transmission over a LOS path or over one via a strong specular reflection. Thus, to be of value, FD-MIMO is best implemented with mMIMO in the millimeter wave bands.

FD-MIMO is realized by employing *active antenna system* (AAS) technology. With AAS, each array input is integrated with a separate RF transceiver unit. AAS allows the phase and amplitude of the signals to each antenna input to be controlled electronically, resulting in very flexible beamforming. Specifically, it allows multiple beams to be generated, with each individually directable in three dimensions.

In smaller cells such as are found with millimeter wave networks, FD-MIMO is particularly applicable. This is because in such situations the vertical scale is likely to be somewhat comparable with the horizontal one. Figure 8.19 depicts such a situation, very likely to be found in dense urban areas, where three users each require different focusing in all three dimensions. Clearly a lack of control here in the vertical plane would lead to sub optimal coverage.

8.3.9 Distributed MIMO

With distributed MIMO (DMIMO) [5], a UE can receive user data from multiple transmission channels simultaneously, each channel transmitted from a spatially different node. In other words, the received MIMO layers are received from

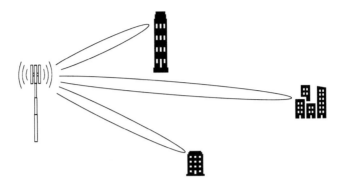

Fig. 8.19 Three-dimensional beamforming

different sites as opposed to SU-MIMO, where all layers are transmitted from the same site. With DMIMO, a large number of neighboring nodes opportunistically form a virtual antenna array for both transmission and reception.

With DMIMO, the signals emanating from the various antennas need to be phase synchronized. This calls for the oscillators at the various nodes to be synchronized in both frequency and phase. One approach to solving this problem is for all the participating nodes to be synchronized to a common beacon. Another requirement is for each node to have access to the same data.

Some of the advantages of DMIMO are its ability to increase capacity, cell-edge throughput, and coverage through the spatial multiplexing of uncorrelated signals where large-scale fading, if experienced, is independent. Some disadvantages are the increased system complexity, for example, the oscillator synchronization issue discussed above, and the large signaling overhead required.

8.3.10   User Equipment Antennas

At sub 6 GHz frequencies, the carrier wavelength is of the same order of the size of the UE and as a result antenna accommodation is typically limited to omnidirectional, dual port, or four port antennas. At millimeter wave frequencies, however, the wavelength is small compared to the size of the UE. This makes it possible to accommodate beamforming antenna arrays with many elements, which is desirable, as this permits relatively large gain that's helpful in overcoming the large path loss. Since the transmission signal can impinge on the UE from any of its sides, it's desirable to have omnidirectional coverage. One way to accomplish this is to have two or more flat arrays that cover different angular sectors, with each array being steerable. With proper placement it should be possible to always have the antenna near optimally directed even when moving.

A complicating factor in UE antenna design is the fact that several millimeter wave frequency bands have to be supported, with the bands differing significantly in frequency. Since array antenna design is wavelength sensitive, many different array designs may need to be incorporated.

8.3.11 Beam Management

In NR, analog or hybrid beamforming is expected to dominate at millimeter wave frequencies, where beamforming is used to establish highly directional transmission links. To support this beamforming at both the BS and UE, a set of procedures has been defined in the 3GPP specifications that supports both DL and UL communication. These procedures are referred to as *beam management*. The essential task of beam management is the establishment and retention of an effective beam pair, that is, a transmitter created beam and a receiver created beam, each beam pointing in a direction such that together they produce optimum connectivity. This connection can be line of sight, and thus the beams point toward each other. It could be, however, that the line of sight is obstructed, but a specular reflection point exists such that with each beam pointing at the reflection point, a good propagation path is established. With NR, beam management is used for both the data and control channels.

In many situations, the optimum DL transmitter/receiver beam pair may also be the optimum UL transmitter/receiver pair. This would obviously be the case with TDD. In 3GPP, this reciprocity is referred to as *beam correspondence*. With beam correspondence a beam pair established for one direction is directly applicable for the other direction. Because beam management is not intended to operate in situations where UE movement is fast and where there is much frequency selective channel variations, beam correspondence can also be established where the duplexing is FDD.

NR beam management can be broken down into five main procedures, namely, *beam sweeping, beam measurement, beam determination, beam reporting,* and *beam recovery*. The following are brief descriptions of these procedures:

- Beam sweeping: The process of covering a defined spatial area via the sequential transmission of a set of analog beams, these beams being transmitted in bursts over a regular interval in a predetermined way, each beam containing a specific reference signal. To facilitate this action, both the base station and UEs have a predefined codebook of directions that cover the defined spatial area. An illustration of beam sweeping at a BS and a UE is shown in Fig. 8.20. The sweeping at the BS and UE is not coincident, and the likely beams chosen for communication following sweeping are shown shaded.
- Beam measurement: Evaluation at the UE or the BS of the quality of the received reference signal from all swept beams. Different metrics may be used in determining signal quality, for example, SNR.
- Beam determination: Selection by the BS or UE of its own Tx/Rx beam(s), based on the information obtained via the beam measurement procedure. The beam

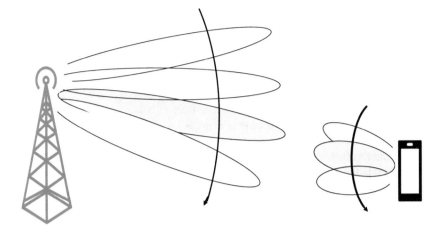

Fig. 8.20 Beam sweeping

selected is that via which the BS or UE experienced the maximum signal quality, if above a predefined threshold. Following selection, a link is established via back and forth communication between the UE and BS. The BS and UE then monitor the quality of the selected beam pair along with other beam pairs. Should the quality of the current beam pair degrade, then there is beam adjustment, where the BS and UE switch to a different beam pair that provides better quality. Beam adjustment may also include narrowing the shape of the beam to improve link quality. Beam adjustment is carried out for DL and UL transmission directions. However, if there is beam correspondence, then beam adjustment need only be carried out in one of the directions, say from the BS in the DL direction, as it can be assumed that the adjusted pair is correct for the opposite direction.

- Beam reporting: Reporting by the UE back to the BS information as to which DL beam or beams have the highest quality and what beam decision has been taken.
- Beam recovery: A procedure implemented when the existing link between the BS and the UE fails, due, for example, to blockage, and thus needs to be reestablished. Failure here is defined as when the BER exceeds a certain value. Following failure, the UE attempts to identify a new beam pair via which connectivity may be reestablished. Upon identification, it sends a beam recovery request to the network which initiates a new access procedure.

8.4   5G NR Multiple Antenna Options

Multi-antenna transmission options specified for 5G NR build on, and in some cases replace, those options specified for LTE networks [6]. The main driver for advanced techniques is the inclusion of millimeter wave communications in 5G systems which brings with it higher propagation losses and increased thermal noise power in cases of wider bandwidths. The key addition to address this change is steerable beamforming via antenna arrays with a very large number of elements which provide high gain and hence allows extended coverage. At the sub 6 GHz bands, the main improvement is mMIMO coupled with more focused beamforming. This combination, via improved spatial separation and thus minimized intercell interference, improves cell spectral efficiency by allowing an increase in the number of simultaneous full resource users. Note, however, that at these lower frequencies, the physical size may limit the number of elements in an antenna array since the area occupied by each element is proportional to the square of the wavelength.

MIMO transmission and beamforming of both control and traffic channels is a distinguishing feature of 5G NR. In LTE, the DL control channels use transmit diversity to ensure sufficient link budget, while the traffic channels use spatial multiplexing MIMO. With NR, the control channels rely on beamforming to achieve coverage. In sub 6 GHz bands, transmit diversity would probably have proved adequate in NR as it has in LTE. However, as discussed above, at millimeter wave bands beamforming transmission is necessary to achieve reliable coverage, thus the pragmatic decision to use beamformed control channels throughout. Note, however, that NR supports specification transparent transmit diversity schemes in the UL. Thus, if the UE is equipped with multiple transmit antennas, schemes such as CDD, which require no change to the receiver processing, may be designed in by the equipment provided.

For the DL, codebook-based closed loop precoding is specified to support up to eight layers of SU-MIMO and up to twelve layers of MU-MIMO. The transmitter may use a different precoder matrix in different parts of the transmission bandwidth, thus resulting in frequency selective precoding. This should enhance performance in millimeter wave bands where bandwidths are likely to be large and thus exhibit channel variability across the band.

For the UL, two modes of transmission are specified, namely, codebook based and non-codebook based. In codebook-based transmission, the BS provides the UE with a transmit precoding matrix indication. The UE uses this indication to select the main UL data channel precoder from the codebook. In non-codebook-based transmission, which requires beam reciprocity, the precoder is determined locally based on received DL reference signals. Up to four layers of SU-MIMO are supported when transmission is via OFDM-CP waveform. However, when transmission is via DFTs-OFDM waveform, only single-layer transmission is supported.

Full-dimension MIMO (FD-MIMO) is supported in 5G NR. NR plans to support distributed MIMO, but support was not complete in release 15.

8.5 Summary

In this chapter we reviewed, at a high level, how multiple antenna techniques, which aid in the increase of capacity and coverage, are applied in 5G NR. Common to all such techniques is the use of multiple antennas at the transmitter, at the receiver, or both, together with intelligent signal processing and coding. Specifically, we reviewed:

- Spatial diversity (SD), which provides protection against deep fading by combing signals which are unlikely to suffer deep fades at the same time.
- Spatial multiplexing multiple-input, multiple-output (SM-MIMO), which permits, in general, the transmission of multiple data streams using the same time/frequency resource thus improving spectral efficiency.
- Beamforming, which at the transmit end permits the sharp focusing of the antenna beam, thus increasing transmit antenna gain in a given direction, and which at the receive end also permits, but to a lesser extent, the focusing of the antenna beam, hence increasing receive antenna gain in a given direction.

Multiple antenna techniques, though widely used in 3GPPs 4G LTE, have been deployed in 5G NR with greater complexity but improved results. As an example, beamforming, though utilized in LTE to a limited extend, now forms the basis of mmWave transmission. Going forward, one can expect to see more and more innovations in this arena as the push to increase spectral efficiency and capacity continues.

References

1. Alamouti SM (1998) A simple transmit diversity technique for wireless communications. IEEE J Sel Areas Commun 16(8):1451–1458
2. Marzetta TL (2010) Noncooperative cellular wireless with unlimited numbers of base station antennas. IEEE Trans Wirel Commun 9(11):3590–3600
3. Gunnarsson S et al (2018) Channel hardening in massive MIMO – a measurement based analysis. arXiv:1804.01690v2, June 2018
4. Ahmadi S (2019) 5G NR; architecture, technology, implementation, and operation of 3GPP new radio standards. Academic Press, London
5. Roh W, Paulraj A (2002) Outage performance of the distributed antenna systems in a composite fading channel. In: Proc. IEEE 56th vehicular technology conference, Vancouver, Canada, Sept. 2002, vol 3, pp 1520–1524
6. Dahlman E, Parkvall S, Skold J (2018) 5G NR: the next generation wireless access technology. Academic Press, London

Chapter 9
Physical Layer Processing Supporting Technologies

9.1 Introduction

In NR, the main downlink physical layer data carrying channel is called the Physical Downlink Shared Channel (PDSCH), and the main uplink physical layer data carrying channel is called the Physical Uplink Shared Channel (PUSCH). Data is presented to these channels in the form of transport blocks. The overall data processing in these channels is largely similar and as shown in Fig. 9.1. Most of the key technologies required in this processing have already been covered in prior chapters of this text, and, where so, the appropriate section is indicated on the figure. The purpose of this chapter is to cover those necessary technologies not yet covered (shown shaded), some simple but key to the process chain, and to the expand discussion on selected ones previously covered. Thus, by the end of this chapter, we will have covered the key technologies necessary to comprehend in some detail 5G NR physical layer processing of the main data carrying channels which will be addressed in Chap. 10.

9.2 LDPC Base Graph Selection and Code-Block Segmentation

In NR, one or two data-carrying transport block(s) of variable size is delivered to the physical layer from the layer above, the Medium Access Control (MAC) layer (the MAC layer will be addressed in Chap. 10). The first action at the physical layer is the attachment, per transport block, of a CRC, CRC_{TB}, say (Sect. 6.2), as shown in Fig. 9.1, thus creating a *code block*. Next, the LDPC base graph to be used is selected based on the transport block size and the coding rate indicated (Sect. 6.3.4.1). Code

© Springer Nature Switzerland AG 2020
D. H. Morais, *Key 5G Physical Layer Technologies*,
https://doi.org/10.1007/978-3-030-51441-9_9

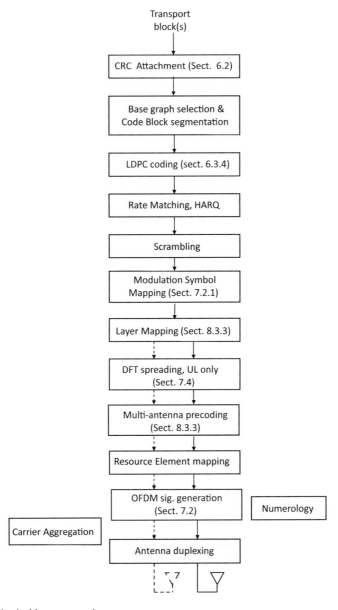

Fig. 9.1 Physical layer processing

block segmentation follows. Here, should a transport block exceed a certain size (8424 bits for LDPC base graph 1, 3824 for base graph 2), the transport block and its attached CRC is broken up into several equal sized code blocks, and each code block has a CRC, CRC_{CB} say, of length 24 bits, attached as shown in Fig. 9.2.

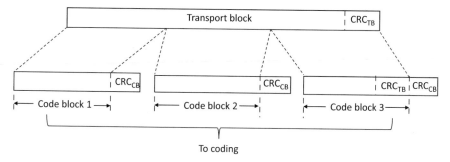

Fig. 9.2 Code-block segmentation

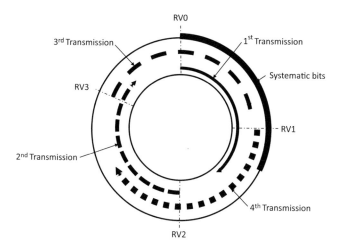

Fig. 9.3 Circular buffer for incremental redundancy

9.3 Rate Matching and HARQ Functionality

Following code-block segmentation, each code block and its attached CRC undergoes LDPC coding (Sect. 6.3.4), followed by rate matching and HARQ functionality [1], the subject of this section.

Rate matching and HARQ functionality, performed separately for each LDPC coded code block and its attached CRC, serve the function of extracting a suitable number of coded bits to match the time/frequency resources assigned for transmission as well as that of generating different redundancy versions needed for effective HARQ operation. Rate matching is performed separately for each code block and attached CRC and consists of bit selection and bit interleaving. The first action taken is to puncture a number of systematic bits, the fraction of bits punctured being up to one-third of the systematic bits, depending on the code-block size. The remaining coded bits are then written into a "circular" buffer, starting with the non-punctured systematic bits and continuing with the parity bits as shown in Fig. 9.3.

As can be seen, such a buffer operates as if it were connected end to end. NR uses four redundancy versions, their locations within the buffer having different starting positions as shown in the Fig. 9.3. The bits chosen for transmission is the result of reading the required number of bits from the circular buffer, the exact set of bits chosen depending on the redundancy version (RV) selected. By selecting different RVs, different sets of coded bits representing the same set of information bits are created, and these bits used when enabling HARQ with incremental redundancy. The starting points on the circular buffer are specified such that both RV0 and RV3 are self-decodable, meaning that they include the systematic bits under typical conditions. It will be noted in Fig. 9.3 that RV3's starting point is located after the 9 o'clock position. This is to allow more systematic bits to be included in the transmission. The default order of the redundancy versions is RV0, RV2, RV3, and RV1. Thus, every second retransmission is typically self-decodable. The length of transmission for each RV is determined by the amount of available transmission resources.

Following rate matching via the circular buffer, the bits generated for each block are interleaved by having the bits from the circular buffer written row-by-row into a block interleaver and read out column-by-column. The final action in rate matching functionality is the concatenation of the code blocks.

9.4 Scrambling

Each block of coded bits created by the HARQ functionality is scrambled. Scrambling is achieved, as shown in Fig. 9.4, by generating a repetitive but long pseudorandom bit sequence and logically combining the generated sequence with incoming data. The scrambled output assumes properties similar to the pseudorandom sequence, irrespective of input data properties. The purpose of scrambling here is intercell interference suppression. With the same spectrum being likely used in all neighboring cells (see Sect. 1.5.2.1), such interference is a real possibility. Without scrambling, the channel decoder could possibly be as equally matched to an interfering signal as to the desired signal using the same frequency resource and thus be unable to properly suppress the interference. Different scrambling sequences are applied in the DL to neighboring BSs and in the UL to different UEs. As a result, interfering signals, after descrambling, are randomized making them appear more noise like and thus more easily addressed via the channel decoding.

Fig. 9.4 Scrambler

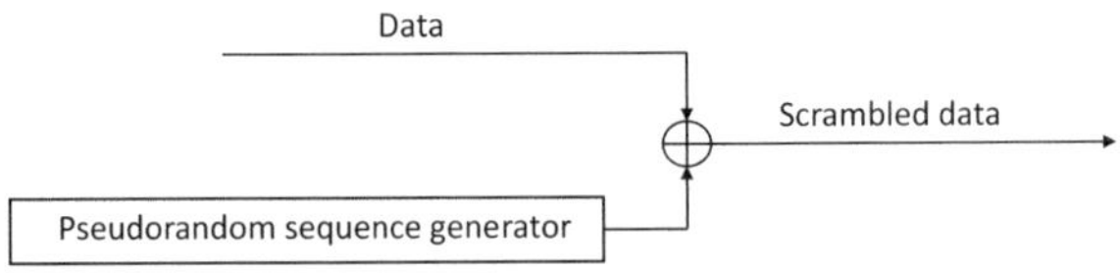

9.5   Resource Element Mapping

Following scrambling, the block of scrambled bits undergoes modulation symbol mapping (Sect. 7.2.1) followed by layer mapping (Sect. 8.3.3). If it is UL processing using optional DFTS-OFDM (Sect. 7.4), then layers received from the layer mapper are DFT spread at this stage. If UL or DL processing is using CP-OFDM, then this stage is skipped. Layers then undergo multi-antenna precoding (Sect. 8.3.3) followed by the resource element mapping stage, the subject of this section.

A resource element is the smallest physical resource in an OFDM time-frequency resource grid. It represents one modulation symbol (time) by one subcarrier (frequency). In NR a *resource block* (RB) is one-dimensional in the frequency domain only and is 12 subcarriers wide. Resource mapping [1] is the taking of modulation symbols to be transmitted on each antenna port and mapping them to the set of resource elements available in a set of resource blocks dictated by the MAC scheduler. It is here that each subcarrier is modulated with a modulation symbol to create a modulated subcarrier.

What the scheduler dictates for transmission is actually a set of virtual resource blocks (VRBs) and a set of OFDM symbols. Within these VRBs, modulation symbols are mapped to resource elements in a frequency first, time second manner, this approach allowing low latency. VRBs containing modulation symbols are mapped to physical resource blocks (PRBs) in one of two methods, namely, non-interleaved mapping and interleaved mapping.

In non-interleaved mapping, the VRBs in a bandwidth section maps directly to the PRBs in the same bandwidth. This method is helpful when the scheduler is trying to allocate resources to that section of the available spectrum with the most favorable channel conditions.

In interleaved mapping, the VRBs are mapped to the PRBs via an interleaver, with output frequencies spanning the whole available bandwidth. This method achieves frequency diversity which is helpful in situations such as (a) low rate services such as VoIP where signaling associated with channel-dependent scheduling could lead to significant relative overhead, or (b) where a UE is travelling at such a speed as to make it difficult to track instantaneous channel conditions and schedule spectrum resources accordingly.

9.6   Numerologies

The output of the resource element mapper drives the OFDM signal generator (Sect. 7.2). In NR there are a number of specification options associated with OFDM-based transmission. Here we examine what is referred to as *numerology*. A numerology (μ) is defined by *subcarrier spacing* (SCS) and cyclic prefix (CP) overhead. In LTE the subcarrier spacing is 15 kHz except for one special case (multicast broadcast single frequency network) where it is 7.5 kHz. In NR several

Table 9.1 NR numerology

Numerology	Subcarrier spacing (kHz)	OFDM useful symbol length (μs)	Cyclic prefix duration (μs)	OFDM symbol length (μs)
0	15	66.67	4.69	71.35
1	30	33.33	2.34	35.68
2	60	16.67	1.17	17.84
2	60	16.67	4.17 (extended)	20.84
3	120	8.33	0.59	8.92
4	240	4.17	0.29	4.46

numerologies are supported which can be mixed and used simultaneously. Multiple subcarrier spacings are derived by scaling a basic subcarrier spacing (15 kHz) by an integer μ. The subcarrier spacing Δf for numerology μ is given by $\Delta f = 15 \times 2^\mu$ kHz. The numerology applied is selectable, independent of the frequency band, but for reasons that will be discussed below, a low subcarrier spacing is not recommended for very high carrier frequencies. Table 9.1 shows supported numerologies in NR. Note that in the table, the OFDM symbol lengths shown are "normal" lengths with "normal" cyclic prefix lengths. Each numerology has two symbols of "long" length per 1 ms subframe (subframe defined in Sect. 10.7), generated by increasing the length of the normal cyclic prefix. This is done to ensure that each numerology has an integer number of symbols within each 0.5 ms time window.

There are many advantages to offering multiple SCS options. Symbol length is inversely proportional to SCS. Thus, with the smaller SCSs, we have the larger symbol lengths. As NR uses the same ratio of CP length to OFDM symbol length, the smaller SCSs have the larger CP lengths and are thus the most tolerable to the effects of multipath delay spread. With the larger SCSs, on the other hand, the symbol durations are shorter. This results in faster transmission turnaround. Also, it results in lower sensitivity to phase noise, since the negative impact of phase noise reduces with subcarrier spacing as discussed in Sect. 7.2.4. Finally, it minimizes the possible degrading effect of Doppler shift. As indicated in Sect. 4.7.5, a signal undergoes Doppler shift-induced fast fading if the modulating signal bandwidth B_s, in this case SCS, is less than the maximum Doppler shift B_d. For negligible impact, we would like B_s to be very much larger than B_d. As shown in Table 4.2, for a carrier frequency of 40 GHz and a speed of 120 km/h, B_d is 8.9 kHz. Thus, under these conditions, we would like SCS to be on the order of 100 kHz. In NR, 15, 30, and 60 kHz SCS values are used for data channels in the below 10 GHz frequency range. These frequencies allow relatively large cell sizes where multipath effects may be large, but Doppler shift and phase noise effects are low, given that Doppler shift is directly proportional to carrier frequency and phase noise is proportional to the square of carrier frequency. For data channels in the millimeter wave bands, however, 60 and 120 kHz SCS are used. Here, due to high propagation loss, cell sizes tend to be small. Thus, multipath effects are low, and therefore a small CP is acceptable. However, as Doppler shift and phase noise are likely to be large, the larger subcarrier spacings are advantageous.

9.7 Transmission Signal Duplexing

In Fig. 9.1, for purposes of simplicity, up-conversion is omitted. In real systems, however, baseband OFDM signals are up-converted to RF (Sect. 7.2.1), where prior to transmission they undergo antenna duplexing so that one antenna can serve both transmission and reception. In this section we review antenna-duplexing schemes.

A transmission signal duplexing scheme is a method of accommodating the bidirectional communication of signals between two devices. In mobile access systems, this means the bidirectional communication between the BS and a UE. There are three duplexing schemes that are utilized in mobile access networks, namely, Frequency Division Duplexing, Frequency Switched Division Duplexing, and Time Division Duplexing.

Frequency Division Duplexing (FDD) is the traditional form of duplexing. In PMP systems utilizing FDD, the DL and UL frequency channels are separate, and thus all UEs can transmit and receive simultaneously. The channels bandwidths are usually, but not necessarily, of equal size. For proper operation of FDD is necessary that there be a large separation between the two assigned frequencies. This is because, at a given terminal, BS or UE, it is necessary that the receiver be able to filter out the transmit signal sufficiently to prevent it from interfering with the received signal.

Frequency Switched Division Duplexing (FSDD), also known as *half-duplex-Frequency Division Duplexing* (*H-FDD*), is a duplexing scheme in which, like FDD, the DL and UL frequency channels are separate but where some or all of the UEs cannot transmit and receive simultaneously but must do so sequentially. Those UEs that can operate simultaneously are said to operate in a *full-duplex* mode, whereas those that must operate sequentially are said to operate in a *half-duplex* mode. The design of UEs that operate in a half-duplex mode is simplified as the antenna-coupling device now consists essentially of only a switch that switches between the transmitter output and receiver input instead of a frequency-separating device. An example of time allocation on a FSDD system is shown in Fig. 9.5.

Time Division Duplexing (TDD) is a duplexing scheme where the downstream and upstream transmissions share the same frequency channel, but do not transmit simultaneously. However, because of the rapid speed of switching between these

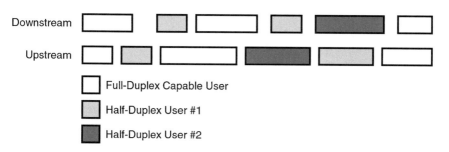

Fig. 9.5 Example full-duplex and half-duplex time allocation

transmissions, "simultaneous" two-way communication is preserved. Whereas FDD transmission requires a large frequency separation, TDD requires a guard interval between transmission and reception. This interval must be large enough to allow for, among other factors, a transmitted signal to arrive at its intended terminal before a transmission at that terminal is started, thus switching off reception. The guard interval is typically on the order of 100 to a few hundred microseconds, with the higher end values being required for cells covering large distances. Note, however, that large guard intervals reduce system capacity as the guard time is not available for useful transmission. The allocation of time between downstream and upstream traffic is normally adaptive, making this technique highly attractive for situations where the ratio of downstream to upstream traffic is likely to be asymmetric and highly variable. A significant advantage of TDD over FDD is channel reciprocity, i.e., both the DL and UL channels share the same propagation characteristics. As a result, the BS is able to estimate the DL channel from its measurement of the UL channel, removing the need to feed this information from the UE up to the BS. One impediment to the deployment of TDD systems is that many global spectrum allocations dictate FDD and prohibit TDD because of the difficulty in coordinating it with FDD systems from an interference point of view. This difficulty arises because the transmission of the same frequency in both DL and UL directions makes the discrimination of a nearby antenna to these signals limited to non-existent.

Where both transmitter and receiver share the same antenna, FDD *antenna coupling* as applied in mobile access systems takes the form of an *antenna duplexer*, as shown in Fig. 9.6a. A key component of the duplexer shown is the antenna circulator. A basic circulator is a three-port device, constructed from ferrite material, with behavior such that an input signal to any port circulates unidirectionally and exits at the next port on the unidirectional path. Thus, in the figure, the transmit signal of frequency f_1 that enters port 1 of the antenna circulator exits port 2, and proceeds via transmission line to the antenna. Likewise, the signal received by the antenna of frequency f_2 enters port 2, exits port 3, and proceeds to the receiver. Because the transmit signal can never be totally absorbed by the antenna system, a small fraction

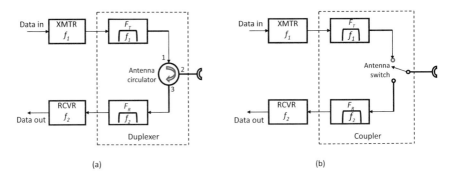

Fig. 9.6 (**a**) FDD antenna duplexing and (**b**) TDD antenna coupling

of it returns to port 2 and ends up at the receiver input. The duplexer filter, F_{T1}, in the transmit leg, limits the level of out of band spurious emissions within regulated levels. It must also limit those emissions that fall within the receiver passband to a degree that any such unwanted input that appears at the receiver front end is at a level low enough as to not degrade receiver BER performance. The filter F_{R2} in the receive leg of the duplexer reduces out of band-interfering signals. It also helps, however, to ensure that the level of the transmitted signal of frequency f_1 reaching the receiver front end does not overload it, resulting in non-linear behavior that degrades the BER performance. Further, in the receiver following F_{R2}, the signal of frequency f_1 is seen as noise and thus must be further filtered to a level below the receiver sensitivity.

In TDD, where both transmitter and receiver share a common antenna, antenna coupling takes the form shown in Fig. 9.6b. The primary difference between this coupling and that employed for FDD is the replacement of the circulator with a switch. Switching is normally relatively fast, and the switching time plus the time for circuitry to switch from DL to UL is typically on the order of 10–20 μs. Thus, in most situations, it does not measurable increase the required guard time. Filters are shown in the transmit and receive legs but serve somewhat different purposes to those shown in the FDD duplexer. On the transmit side, the filter F_{T1} serves, as with FDD, to reduce out of band emissions to within regulated levels. There is no need, however, to be concerned about the transmit signal level into the receiver as during transmission the switch isolates this signal from the receiver. On the receive side, the filter F_{R2} serves only to reduce out of band emissions, not the level of the transmit signal. The cost of implementing a TDD antenna coupler is lower than that of implementing an FDD one as the cost of the switch is lower than that of a circulator. A small disadvantage of TDD relative to FDD, however, is its higher latency as data may not be instantly routed for transmission but must wait the assigned time slots. Typically, this delay may be on the order of a few milliseconds.

The duplexing scheme applied is dictated by the allocation of spectrum by the governing authority. Allocation is either paired, thus facilitating FDD, or unpaired, facilitating TDD. In general, bands below about 10 GHz tend to be, but not always, paired. In the millimeter wave range, on the other hand, assigned bands are increasingly unpaired. 5G NR can operate in both FDD and TDD mode. FDD will be the main duplexing scheme in the below 10 GHz bands and TDD more common in the millimeter wave bands. Because of their high signal loss, systems operating in the millimeter wave range will typically cover small cells with a relatively small number of users per BS. In such situations the DL/UL traffic requirements can vary rapidly. To make optimum use of available total DL/UL capacity, NR uses *dynamic TDD* which allows the dynamic assignment of DL and UL resources, i.e., the ratio of time assigned to DL traffic to that assigned to UL traffic varies dynamically according to need.

9.8 Carrier Aggregation

Carrier aggregation (CA) [1, 2] refers to the concatenation of multiple carriers that can be transmitted in parallel to and from the same device thus increasing the available transmission bandwidth and in turn maximum data rate achievable. Aggregated carriers are called component carriers. As shown in Fig. 9.7, component carriers need not be contiguous and in the same frequency band but can be aggregated in one of three scenarios:

- Intra-band aggregation with contiguous component carriers
- Intra-band aggregation with non-contiguous component carriers
- Inter-band aggregation with non-contiguous component carriers

5G NR supports carrier aggregation of up to 16 component carriers, in both the below 10 GHz range and the millimeter wave range, component carriers having the same or different numerologies, and possibly of different bandwidths and duplexing schemes. As the maximum NR channel bandwidth is 400 MHz, the maximum aggregated bandwidth possible, but highly unlikely, is 16×0.4 GHz = 6.4 GHz!

The number of component carriers in any aggregation is independently configured for DL and UL. In the inter-band aggregation scenario where there are multiple TDD carriers, the transmission direction of the different carriers need not be the same. This implies the possibility of a TDD device capable of carrier aggregation transmitting on one frequency while receiving on another and thus needing an RF duplexer.

In the NR specification, each component carrier appears as a separate "cell," and a carrier aggregation capable UE is said to be able to communicate with multiple

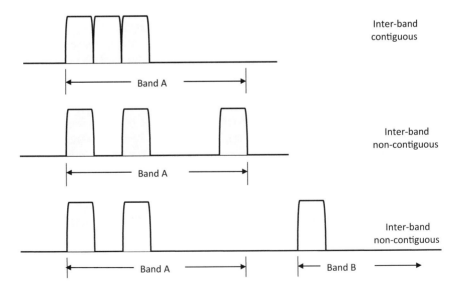

Fig. 9.7 Types of carrier aggregation

cells. Such a UE connects to one primary serving cell (PCell) and one or more secondary serving cells (SCells). The PCell is the cell that the UE initially selects and is connected to. Once connected to the BS via the PCell, one or more SCells can be configured. Once configured, secondary cells can be rapidly activated or deactivated based on need.

A single scheduler entity in the BS MAC schedules all UEs and all their corresponding component carriers. At the physical layer, as depicted in Fig. 9.1, each transport block is mapped to a single component carrier. Even when multiple component carriers are scheduled on a UE simultaneously, coding, HARQ, modulation, numerology, and resource element mapping, along with corresponding signaling, are performed independently on each component carrier.

9.9 Summary

In NR, the main downlink physical layer data carrying channel is the Physical Downlink Shared Channel (PDSCH) and the main uplink physical layer data carrying channel is the Physical Uplink Shared Channel (PUSCH). The overall data processing for these channels is largely similar. Most of the key technologies required in this processing were covered in earlier chapters. In this chapter, those necessary technologies not yet covered were discussed, some simple but key to the processing chain, and the dialogue expanded on selected ones previously covered. Thus, we have now covered the key technologies necessary to comprehend in some detail 5G NR physical layer processing of the main data carrying channels. This will be addressed in Chap. 10 along with a description of the overall physical layer.

References

1. Dahlman E, Parkvall S, Skold J (2018) 5G NR: the next generation wireless access technology. Academic, London
2. Ahmadi S (2019) 5G NR; architecture, technology, implementation, and operation of 3GPP new radio standards. Academic, London

Chapter 10
5G NR Overview and Physical Layer

10.1 Introduction

This text deals primarily with the physical layer. However, to better understand its operation, it is helpful to stand back a bit and study, at a high level, its surrounding architecture and protocols. To this end, we review in this chapter the main architecture options for connection to the core network followed by the *Radio Access Network* (RAN) protocol architecture. The RAN is responsible for all radio-related functions such as coding, modulation, HARQ operation, physical transmission, scheduling, etc. Following this, we narrow our study within the RAN to the physical layer, with emphasis on how physical channels are structured, be they user data conveying or control data conveying, as well as how physical signals are created, such signals being those originating solely in the physical layer. Procedures such as initial access, scheduling, and uplink power and timing control are introduced. How maximum user data rates and low latency are achieved is demonstrated, 5G operating frequency spectrum reviewed, some typical base station and UE parameters presented, and relevant 3GPP specifications listed.

10.2 Connection to the Core Network

The *core network* is responsible for the overall control of the UE. Functionally, it sits above the RAN and handles functions not related to radio access but required for the providing of a complete network such as authentication and the establishment, maintenance, and release of communication links. To ease the transition to a fully independent 5G network, 3GPP specified two primary core connection architectures: *non-standalone*, for which there are a number of variations based on the routing of user data, and *standalone*. Shown in Fig. 10.1 is the stand-alone architecture called Solution 2 and one of the non-stand-alone architectures called Solution 3x.

© Springer Nature Switzerland AG 2020
D. H. Morais, *Key 5G Physical Layer Technologies*,
https://doi.org/10.1007/978-3-030-51441-9_10

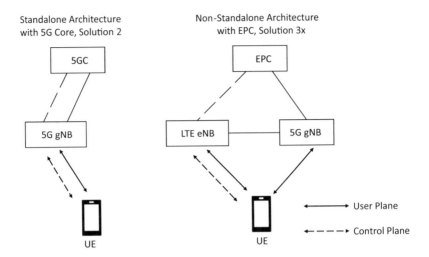

Fig. 10.1 Non-stand-alone and stand-alone core connections

In the non-stand-alone architecture, both the LTE core, called the Evolved Packet Core (EPC), and the LTE base station, called the eNodeB, are utilized. Control plane data is all done via connection to the eNodeB, and user plane data to/from the UE is via either the eNodeB or the NR base station called the gNB or simultaneously with both. When the latter is the case, this is referred to as *dual connectivity*, which can increase the user data rate and increase reliability. From the point of view of available 5G service, only enhanced mobile broadband (eMBB) is available with the non-stand-alone architecture. Note, however, that dual connectivity can also be 5G–5G, where the UE communicates with two gNBs. 3GPP Release 15 assumes synchronization between the different gNBs; however, in Release 16, this limitation is removed.

In the stand-alone architecture, no elements of LTE are utilized. The UE is connected to the 5G core (5GC) via the gNB. This makes available all new functionality provided by the 5GC. Thus, in addition to eMBB, ultra-reliable low-latency communication (URLLC) is available via Rel. 15 and massive machine-type communications (mMTC) via Rel. 16.

10.3 RAN Protocol Architecture

The 5G RAN protocol architecture of the gNB and the UE [1–3] is shown in Fig. 10.2. The protocols consist of a *user plane* (UP) and a *control plane* (CP). The UP transports user data which enters/leaves from above in the form of IP packets. The control plane transports control signaling information and is mainly responsible for connection establishment and maintenance, mobility, and security. The protocol

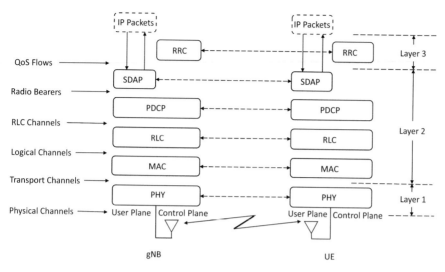

Fig. 10.2 User plane and control plane protocols

stack is divided into layers. Layer 1 is the physical layer. Layer 2 encompassed the MAC, RLC, PDCP, and SDAP sublayers. Layer 3 is the RRC layer. It will be noted that most of the protocols are common to the user plane and control plane. Shown in Fig. 10.2 is the nomenclature of connections between layer/sublayer and sublayer/sublayer. Logical channels, which are offered to the RLC sublayer by the MAC sublayer, are characterized by the type of information they carry. Transport channels, which are offered to the MAC sublayer by the physical layer, are characterized by how and with what characteristic information is transferred over the radio interface. Data on a transport channel is arranged into *transport blocks* (TBs). Physical channels carry data from higher layers between the gNB and UE.

Before proceeding further, the definition of some commonly used terms in describing the various protocols is in order:

- A *data unit* is the basic unit exchanged between different layers of a protocol stack.
- A *service data unit* (SDU) is a data unit passed by a layer above to the current layer for transmission using services of the current layer.
- A *protocol data unit* (PDU) is a data unit created by the current layer via the adding of a header to the received SDU prior to transportation to the layer below. The header added describes the processing carried out by the current layer.
- A *data radio bearer* (DRB) is a radio bearer conveying user data received from the SDAP.
- A *signaling radio bearer* (SRB) is a radio bearer conveying control information received from the RRC.

10.4   Layer 3 (RRC) Description

As seen in Fig. 10.2, Level 3 has one protocol, namely, the *Radio Resource Control* (RRC) protocol, which resides in the control plane. The overall task of the RRC is to configure the UE with the parameters required by the other protocol layers to establish and maintain connectivity between the UE and the gNB. It does this by handling control layer procedures, including:

- Broadcast of *system information* (SI). System information is all the non-device-specific information that a UE needs to be able to communicate with the network via the gNB. System information consists of a *master information block* (MIB) and a number of *system information blocks* (SIBs).
- Transmission of paging messages emanating from the 5GC or the gNB to the UE via the gNB to notify the UE about connection requests.
- The establishment, maintenance, and release of an RRC connection between the UE and the gNB.
- Security functions.
- Establishment, configuration, maintenance, and release of SRBs and DRBs.
- Mobility functions, including UE cell selection and reselection and handover.
- QoS management functions.
- UE measurement reporting.
- Detection and recovery from a failure of the radio link.

RRC messages are sent via SRBs as these always have a higher priority than DRBs, thus ensuring that they are sent as quickly as possible.

The RRC can be in one of three different states depending on the traffic activity. These states are RRC_IDLE, RRC_INACTIVE, and RRC_CONNECTED.

In RRC_IDLE state, the parameters necessary for communication between a UE and a gNB have not been assigned to a specific gNB. Thus, no data transfer can take place to/from the UE, and the UE "sleeps" most of the time, waking up only periodically to see if there is a paging message from the network which may lead to a change of RRC state.

In the RRC_INACTIVE state, the parameters necessary for communication between the UE and a specific gNB have been assigned to a specific gNB, but there is no data transfer. However, when there is data to be transferred, the RRC quickly changes state to RRC_CONNECTED.

In the RRC_CONNECTED state, the parameters necessary for communication between the UE and a specific gNB have been assigned to a specific gNB, and data transfer takes place.

10.5   Layer 2 User Plane and Control Plane Protocol Description

A detailed layer 2 user plane protocol and control plane structure is shown in Fig. 10.3.

Following is a description of the various protocol sublayers, starting from the top.

10.5.1   The Service Data Adaptation Protocol (SDAP) Sublayer

In NR, IP packets (IPv4 or IPv6) as described in Sect. 2.2 are mapped from the 5G core network to data radio bearers (DRBs) according to their quality of service (QoS) requirements. The *Service Data Adaptation Protocol* (SDAP) is a user plane sublayer and is responsible for this mapping. It also marks both the DL and UL packets with an identifier as to the specified QoS. The SDAP is a new protocol and not present in 4G. If the gNB is connected to the EPC, as in the non-stand-alone case shown in Fig. 10.1, the SDAP is not used.

10.5.2   The Packet Data Convergence Protocol (PDCP) Sublayer

The *Packet Data Convergence Protocol* (PDCP) is responsible for several services and functions [1]. The following is a description of the main ones:

- Sequence numbering, in the downward flow in the user plane, of data radio bearers so that in-sequence reordering is possible in the upward flow.
- Header compression, in the user plane only, to reduce the number of bits transmitted over the radio interface. Header compression is accomplished via robust header compression (ROHC) as discussed in Sect. 2.3. In the upward flow, header decompression is applied.
- Integrity protection, in the downward flow in the control plane, to ensure that control messages originate from the correct source. In the upward flow, integrity verification is applied.
- Ciphering, also known as encryption, in the downward flow in both the user plane and control plane, to ensure that intruders cannot access the data and signaling messages that the UE and the gNB exchange. In the upward flow, deciphering is applied.
- Duplicate detection and removal in the upward flow in the user plane and, if in order delivery to layers above is required, reordering to provide in-sequence delivery.

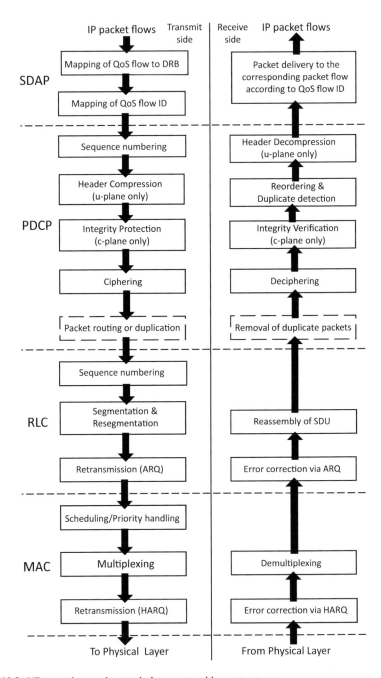

Fig. 10.3 NR user plane and control plane protocol layer structure

10.5.3 The Radio Link Control (RLC) Sublayer

The RLC sits below the PDCP sublayer and interfaces with the MAC sublayer below via logical channels. Depending of the class of service to be provided, the *Radio Link Control* (RLC) sublayer can be structured in one of three transmission modes, namely, the *transparent mode* (TM), the *unacknowledged mode* (UM), and the *acknowledged mode* (AM). In the transparent mode, data unit flow is transparent, and no headers are added. In the unacknowledged mode, segmentation (discussed below) is supported, and, in the acknowledged mode, segmentation and retransmission of erroneous packets are supported. The RLC is responsible for several functions and services including:

- Sequence numbering of user data in the downward flow, independent of the one in the PDCP above. This numbering is in support of HARQ retransmissions. In the upward flow, RLC does not facilitate in-sequence delivery, minimizing the overall latency. This task is left to the PDCP sublayer above.
- Segmentation in the downward flow, of RLC SDUs, received from the PDCP above, into suitably sized RLC PDUs. In the upward flow, reassembly of the original RLC SDUs.
- Retransmission, in the acknowledgment mode, in the downward flow, via HARQ, of erroneously received RLC PDUs at the receiving end.
- Error correction in the upward flow via HARQ.

10.5.4 The Medium Access Control (MAC) Sublayer

The *Medium Access Control* (MAC) sublayer is the lowest sublayer of Level 2 and interfaces with the physical layer below via transport channels. As indicated above, data on a transport channel is arranged into transport blocks (TBs), and the transmission time of each transport block is called the *Transmission Time Interval* (TTI). One transport block of variable size can be transmitted in each TTI over the radio interface to/from an EU except when there is spatial multiplexing of more than four layers, in which case two transport blocks are transmitted per TTI.

The MAC sublayer is responsible for several functions and services including:

- Mapping between logical channels and transport channels.
- In the downward flow, multiplexing of MAC SDUs (RLC PDUs) belonging to one or different logical channels into transport blocks delivered to the physical layer on transport channels. Associated with each transport block is a *transport format* (TF) which defines how the transport block is to be transmitted over the radio interface. The TF conveys information about the size of the transport block, the coding and modulation scheme to be employed, and the antenna mapping.

– In the upward flow, demultiplexing of MAC SDUs belonging to one or different logical channels from transport blocks delivered from the physical layer on transport channels.
– Error correction via HARQ.
– Scheduling and scheduling-associated functions.

The MAC sublayer provides services to the RLC via logical channels that can be classified into two groups: control channels used for transporting control plane signaling and traffic channels used for carrying user plane data. The logical channels defined in 3GPP NR specifications are:

– The *Broadcast Control Channel* (BCCH): A DL channel used for broadcasting system information (SI) received from the RRC above to all UEs in a cell. In order to access the network, a UE needs to obtain the SI to learn how the system is configured and how to operate within the cell. As indicated above, SI consists of an MIB and several SIBs. The MIB is passed down to the BCH transport channel and the SIBs to the DL-SCH transport channel (transport channels described below). Among the SIBs is SIB1, referred to as *remaining minimum system information* (RMSI), and consists of the system information that the UE needs over and above the MIB in order to access the network.
– The *Paging Control Channel* (PCCH): A DL channel transmitted in multiple cells to page UEs whose location in a given cell is not known to the network and to notify UEs of system information changes and Public Warning System broadcasts.
– The *Common Control Channel* (CCCH): A DL and UL channel used for transmitting control information between UEs and the network for UEs having no RRC connection with the network.
– The *Dedicated Control Channel* (DCCH): A point-to-point DL and UL channel used to transmit dedicated control information between a UE and the network. It is used by UEs having an RRC connection for the configuration of the UE.
– The *Dedicated Traffic Channel* (DTCH): A point-to-point DL and UL channel, dedicated to one UE, for the transfer of user information to/from the UE.

Logical channels are mapped to transport channels. The transport channels defined in NR specifications are:

– The *Broadcast Channel* (BCH): A DL channel broadcast in the entire coverage area of the cell. It conveys the master information block (MIB) received from the BCCH logical channel above. The MIB is a block of data that contains a limited amount of the total information that the UE requires in order to acquire the remaining system information (SIBs) broadcast by the network via the DL-SCH. The transport block that it delivers to the physical layer has a fixed format defined by the specifications.
– The *Paging Channel* (PCH): A DL channel broadcast in the entire coverage area of the cell, used for the transmission of paging information from the PCCH logical channel above. It supports UE *discontinuous reception* (DRX) to enable UE

power saving by sleeping and waking up only at predefined times to receive the PCH.

- The *Downlink Shared Channel* (DL-SCH): The DL channel used to for the transmission of user data received from the DTCH above. It is also used for transmission of SIBs received from the BCCH above, transmission of UE-specific control information received from the DCCH above, and transmission of common control information received from the CCCH. It supports HARQ with soft combining and dynamic link adaptation via the varying of coding, modulation, and transmit power. It further supports both dynamic and semi-static resource allocation, discontinuous reception (DRX), and the possibility to use beamforming.

- The *Uplink Shared Channel* (UL-SCH): Similar to the DL-SCH, the UL-SCH is the UL channel used for the transmission of user data. It supports HARQ with soft combining and dynamic link adaptation via the varying of coding, modulation, and transmit power. It further supports both dynamic and semi-static resource allocation and the possibility to use beamforming.

- The *Random-Access Channel* (RACH): A UL channel used by the UE to request access to the network when the UE does not have accurate UL timing synchronization or does not have any allocated UL transmission resource. It carries the risk of collision with other transmissions in which event it must back off and try again. Note that although it is defined as a transport channel, it does not carry transport blocks.

The mapping between logical channels and transport channels as well as that between transport channels and physical channels is shown Fig. 10.4 for both the DL and UL. Also shown is physical signals which are described below.

An example of Layer 2 user plane downward data flow is shown in Fig. 10.5. Here, at the top, are three IP packets, two on radio bearer RB_x and one on radio

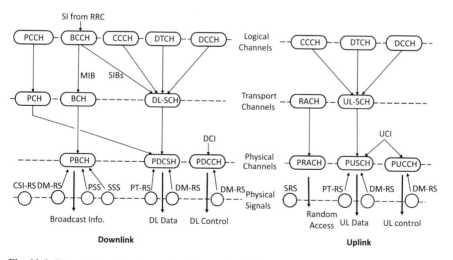

Fig. 10.4 Physical signals and mapping between logical, transport, and physical channels

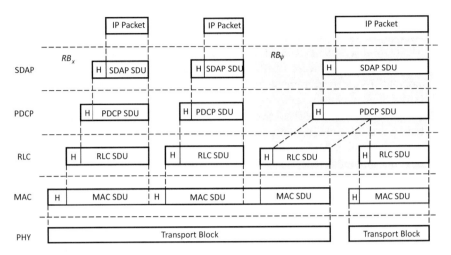

Fig. 10.5 An example of Layer 2 user plane data flow

bearer RB_y. First the IP packets enter the SDAP layer as SDAP SDUs where they each have a header added to create SDAP PDUs. These PDUs enter the PDCP layer as PDCP SDUs where, after processing including header compression, each has a header added to create PDCP PDUs. These PDUs enter the RLC layer where the ones created from radio bearer RB_x become RLC SDUs and the one created from radio bearer RB_y is segmented into two RLC SDU segments. After processing, the segmented and unsegmented SDUs each have a header added to create RLC PDUs which are then forwarded to the MAC layer where they are received as MAC SDUs. These SDUs, after processing, each have a header added creating MAC PDUs. The first three MAC PUDs counting from left to right (two from RB_x and one from RB_y) are multiplexed to form one transport block which is forwarded to the physical layer, and the fourth MAC PDU (from RB_y), on its own, is forwarded to the physical layer as another transport block.

10.6   Layer 1 (Physical Layer) Description

The physical layer, the main focus of this book, is the lowest layer, Layer 1, in the RAN protocol architecture. Its input on the downward flow and output on the upward flow is transport blocks. As mentioned above, one transport block of variable size can be transmitted in each TTI over the radio interface to/from an EU except when there is spatial multiplexing of more than four layers, in which case two transport blocks are transmitted per TTI.

The physical layer is responsible for many functions including CRC attachment, coding, rate matching and HARQ, scrambling, linear modulation, layer mapping, and mapping of the signal to the assigned physical time-frequency resource. It also handles the mapping of transport channels to physical channels. The physical channels defined in NR specifications are:

- The *Physical Downlink Control Channel* (PDCCH): A DL channel used to convey control information to the UE, including scheduling decisions required by the UE to know when and where to receive data, parameters used by the PDSCH for its transmission, and scheduling grants for uplink transmission.
- The *Physical Downlink Shared Channel* (PDSCH): A DL channel that carries user data, system information, paging information, and control information from layers above. Specifically, it provides the physical layer to transport information from the DL-SCH and PCH transport channels. Its allocation and other parameters used for its transmission are signaled to the UE by the PDCCH.
- The *Physical Broadcast Channel* (PBCH): A DL channel fed by the BCH above, which carries some of the required system information to allow a UE to access the network.
- The *Physical Uplink Control Channel* (PUCCH): A UL channel used by the UE to send HARQ acknowledgments to the gNB, indicating whether or not the DL transport block was successfully received, to send reports of the state of the DL channel to aid in the DL channel-dependent scheduling and to request UL scheduling grants.
- The *Physical Uplink Shared* Channel (PUSCH): The UL version of the PDSCH, providing the physical layer to transport user information from the UL-SCH.
- The *Physical Random-Access Channel* (PRACH): A UL channel used to enable the random-access procedure by physically transmitting information from the RACH.

Besides the physical channels, *physical signals* are also employed in NR transmission and their relationship to physical channels shown in Fig. 10.4. Physical signals are time-frequency resources used by the physical layer, but which do not contain information acquired from layers above. They are reference signals (pilot subcarriers) used for different purposes, for example, demodulation and channel estimation, and synchronization signals used for UE synchronization with the gNB. The physical signals defined in NR specifications are:

- *Demodulation Reference Signals* (DM-RSs): Used in both the DL and UL to estimate the radio channel (the channel coefficients) for demodulation purposes. It is UE-specific, confined to a scheduled resource, can be beam-formed, and transmitted only as necessary. In the DL, there is a DM-RS for the PDSCH, one for the PDCCH, and one for the PCBH. In the UL, there is one for the PUSCH and one for the PUCCH.
- The *Phase-Tracking Reference Signal* (PT-RS): Used in both the DL and UL to facilitate compensation of oscillator phase noise (Sect. 7.2.4) and thus employed when transmission is in a millimeter wave band where oscillator phase noise is likely to be high. It allows the reduction of common phase error at the receive end. It is UE-specific, confined in a scheduled resource, and can be beam-formed.

- The *Channel State Information Reference Signal* (CSI-RS): Used in the DL only to allow the UE to acquire channel state information (CSI) and report this to the gNB to facilitate link adaptation. It also supports reference signal received power (RSRP) measurements for mobility and beam management, time/frequency tracking for demodulation, and UL reciprocity-based precoding. It is UE-specific but multiple users can share the same CSI-RS resource. It is defined as either zero power (ZP-CSI-RS) or nonzero power (NZP-CSI-RS). When configured as ZP-CSI-RS, the resource elements assigned to it are made unavailable for PDSCH transmission as they are configured for CSI-RS transmission on another device.
- The *Tracking Reference Signal* (TRS): Used to aid the UE in time and frequency tracking made necessary because of time and frequency variations in its local oscillator. It is not a defined physical signal per se, but rather a resource set consisting of multiple periodic NZP-CSI-RSs.
- The *Sounding Reference Signal* (SRS): Used in the UL only to enable the gNB to perform CSI measurements to be used mainly for frequency domain scheduling and link adaptation. It is also useful in the case of FDD where there is no channel reciprocity and thus no UL CSI information available from UE DL measurements.
- The *Primary Synchronization Signal* (PSS) and the *Secondary Synchronization Signal* (SSS): This pair of signals, used in the DL only, is employed by the UE during initial access (Sect. 10.13). Together, the PSS and the SSS create the capability to transmit one of 1008 possible physical cell identities (cell IDs). In the first stage of initial access, the UE searches for the PSS and SSS in order to detect the presence of a gNB, acquire accurate timing and frequency of the system, and determine the physical cell identity. The PSS provides initial timing and frequency information, and the SSS provides refinement to the initial findings.

10.7   The Frame (Time-Domain) Structure

The NR frame structure supports both frequency division duplexing (FDD) and time division duplexing (TDD) operations in both licensed and unlicensed frequency bands. It facilitates very low latency, fast acknowledgment of HARQ and dynamic TDD. Transmissions in the time domain, in both the DL and UL, are organized into *frames*. There is one set of frames in the UL and one in the DL on a carrier. Each UL frame transmitted from the UE starts at a specified time [4] before the start of its corresponding DL frame at the gNB to ensure that the UL frame, when it arrives at the gNB, is synchronized with the DL frame at the gNB.

Frames are of 10 ms duration, each frame being divided into ten subframes, with each subframe thus being of 1 ms duration. In turn, each subframe is divided into *slots*, each slot consisting of 14 OFDM symbols. Note, however, that there is an exception to this when the subcarrier spacing is 60 kHz and the extended cyclic prefix (Sect. 9.6) is employed. Here there are only 12 OFDM symbols per slot. A

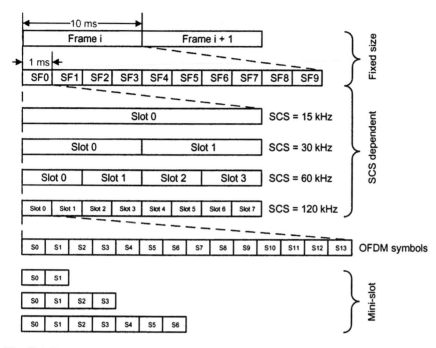

Fig. 10.6 Frame structure and slots. (From [4], with the permission of Elsevier)

slot represents the nominal minimum scheduling interval, referred to as the Transmission Time Interval (TTI), but data can be scheduled to span more than one slot, the latter being referred to as slot aggregation. Recall from Table 9.1 that the OFDM symbol varies with numerology. Thus, the duration of a slot and hence the number of slots per subframe depend on the numerology. This is demonstrated in Fig. 10.6 [4]. The larger the subcarrier spacing, the shorter the slot duration. For subcarrier spacings of 15, 30, 60, and 120 kHz, the associated slot durations are 1, 0.5. 0.25, and 0.125 ms, respectively. In terms of numerology, slot duration is equal to $1/2^{\mu}$ ms. The shorter the slot duration, the lower the latency. This lowered latency is achieved, however, at the expense of a shorter cyclic period which, in a large cell deployment, may not be acceptable due to large rms delay spread. To further aid in latency reduction, reference signals and control signals are located at the beginning of the slot or the set of slots in the case of slot aggregation, as this speeds up the receiver processing.

To allow greater flexibility regarding latency, NR allows using less than 14 symbols for transmission, the resulting smaller effective slots being referred to as *mini-slots*. With the mini-slot structure, 2, 4, or 7 symbols can be allocated with a flexible start position allowing transmission to commence as soon as possible without waiting for the start of a slot boundary. Mini-slots permit shorter latencies even with the 15 kHz subcarrier spacing. There are three good reasons for allowing the use of mini-slots. The first is lowered latency in lower-frequency transmissions where typically 15 or 30 kHz subcarriers are employed. The second is support of analog beamforming in very high-frequency transmission. Here, even though high

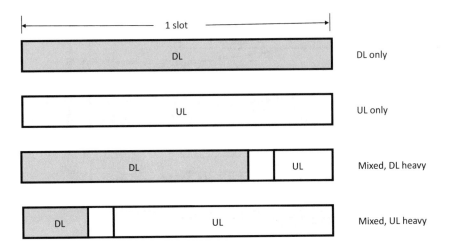

Fig. 10.7 TDD-based frame structure

subcarrier frequencies are used and hence shorter slots, only one beam at a time can be used for transmission, as discussed in Sect. 8.3.7.2, and thus transmission to multiple users must be on a time division basis. Without mini-slots, all 14 symbols would have to be used per UE, before moving on to the next UE. The net effect would be increased multi-transmission latency. With the use of mini-slots, this can be reduced even when transmitting large payloads as with the very large bandwidths available, a few OFDM symbols can be of large data-carrying capacity. The third good reason for employing mini-slots is to facilitate efficient transmission in unlicensed bands. In such bands, the transmitter must constantly monitor the band usage and only transmit when the band is clear. Once it's determined that the band is clear, then the quicker transmission can begin, the better the chance of seizing the spectrum. Mini-slots facilitate this quick action.

In the TDD mode, as shown in Fig. 10.7, a slot can be scheduled for all DL transmission, all UL transmission, or a mixture of both DL and UL, where guard periods are inserted for UL/DL switching.

10.8   The Frequency Domain Structure

In NR, a *resource element*, consisting of one subcarrier during one OFDM symbol, is the smallest physical resource specified. In the frequency domain, 12 consecutive subcarriers of the same spacing are called a *resource block* (RB), the width of a resource block thus being a function of its numerology. The basic scheduling unit is the *physical resource block* (PRB), which is defined as 12 consecutive subcarriers of the same spacing in the frequency domain, i.e., one RB, over one OFDM symbol in the time domain, with all subcarriers within the PRB having the same CP length. Multiple numerologies are supported on the same carrier, and resource block

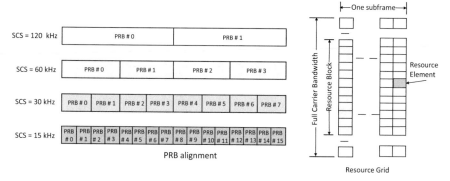

Fig. 10.8 PRB alignment and resource grid

Table 10.1 Minimum and maximum number of resource blocks and corresponding transmission bandwidths

Freq. range	Subcarrier spacing (kHz)	Min (N_{PRB})	Max (N_{PRB})	Minimum bandwidth (MHz)	Maximum bandwidth (MHz)
FR1	15	25	270	4.50	48.60
FR1	30	11	273	3.96	98.20
FR1	60	11	135	7.92	97.20
FR2	60	66	264	47.52	190.08
FR2	120	32	264	46.08	380.16

locations are specified so as to have their boundaries aligned. Thus, as shown in the PRB alignment portion of Fig. 10.8, two PRBs of subcarrier spacing 30 kHz, for example, occupy the identical frequency range as one PRB of 60 kHz subcarriers. Shown in resource grid portion of Fig. 10.8 is a resource element (dark shading) and a PRB (light shading) within a physical resource grid, the latter being described below.

NR specifies minimum and maximum values of RBs (same for DL and UL) per carrier as a function of numerology. Table 10.1 shows these values and their corresponding transmission bandwidths for frequency ranges FR1 and FR2 (Sect. 10.18). For each numerology and carrier, a *resource grid* is defined that covers, in the frequency domain, the full carrier bandwidth being utilized and, in the time domain, one subframe. The resource grids for all subcarrier spacings overlap, and there is one set of resource grids per transmission direction. The resource grid defines the transmitted signal space as seen by the UE for a given subcarrier spacing. However, the UE needs to know where exactly in the available transmitted bandwidth the resource blocks are located. To address this need, NR specifies a common reference point for resource grids referred to as *point A* as well as two classes of resource blocks, namely, *common resource blocks* and *physical resource blocks*. Common resource blocks are numbered from 0 and upward in the frequency domain for a given subcarrier spacing. The center of subcarrier 0 of common resource block 0 for a given subcarrier spacing coincides with "point A." Point A serves as a reference from which the frequency structure can be described and need not be the actual carrier frequency. As a part of the initial access procedure, the location of point A is transmitted to the UE as part the information broadcasted by the PBCH. The physical resource blocks, which indicate the actual transmitted signal spectrum, are then located relative to point A.

10.9   Antenna Ports

In NR, a logical *antenna port* is defined such that the channel over which a symbol on the antenna port is conveyed can be inferred from the channel over which another symbol on the same antenna port is conveyed. In other words, symbols that are transmitted over an antenna port are subject to the same propagation conditions. Thus, the receiving device can assume that two transmitted signals have travelled over the same radio channel and hence experienced the same propagation conditions if and only if they are transmitted from the same antenna port. Demodulation Reference Signals (DM-RSs) are used to help receiving devices estimate the channel that corresponds to the signal transmitted from a specific antenna port. Thus, for the DM-RS associated with the PDSCH, for example, the channel over which a PDSCH symbol on one antenna port is transmitted can be inferred from the channel over which a DM-RS symbol on the same antenna port is transmitted if the two symbols are within the same scheduled PDSCH resource, in the same slot, and undergo the same precoding.

Key to understanding an antenna port is the fact that it doesn't necessarily correspond to a physical antenna. For example, two different signals can be transmitted over several physical antennas via the same pre-antenna paths, as is done with analog beamforming. Nonetheless, the receiver will still see these signals as travelling over the same composite channel and thus emanating from the same antenna port. On the other hand, two different signals transmitted over these same set of antennas but with different precoding for each signal will be seen by the receiver as having been transmitted from two different antenna ports. NR channels/signals and their associated antenna ports are shown in Table 10.2.

10.10   Physical Layer Processing of Transport Channels

The DL transport channels are the DL-SCH, the BCH, and the PCH. The UL ones are the UL-SCH and the RACH. In this section, we examine how transport blocks created by these channels are processed in the physical layer, with the exception of the RACH, which, though defined as a transport channel, does not deliver transport blocks but rather preamble sequences. Time/frequency resource mapping, which can be quite detailed, is only addressed here in the most general terms. For those interested in a more detailed description, it is well covered in [4].

Table 10.2 NR antenna ports

Downlink		Uplink	
Channel/signal	Antenna port starting with	Channel/signal	Antenna port starting with
PDSCH	1000	PUSCH	0
DM-RS for PDSCH	1000	DM-RS for PUSCH	0
PT-RS for PDSCH	1000*	PT-RS for PUSCH	0*
PDCCH	2000	SRS	1000
DM-RS for PDCCH	2000	PUCCH	2000
CSI-RS	3000	DM-RS for PUCCH	2000
TRS	3000	PRACH	4000
PBCH, PSS, SSS	4000		
DM-RS for PBCH	4000		

Note: * signifies lowest number in group

10.10.1 Physical DL Shared Channel (PDSCH) and Physical UL Shared Channel (PUSCH) Processing

Shown in Fig. 10.9 is a block diagram summary of the various processing steps taken on the transmit side by the PDSCH and PUSCH from transport block(s) into the gNB or UE to antenna out (up-conversion is omitted). In the case of the PDSCH, transport blocks are received from the DL-SCH and the PCH. For the PUSCH, transport blocks are received from the UL-SCH. The process shown is the same as that discussed in Chap. 9 and shown in Fig. 9.1 and so requires little additional comment here. Pertinent specifications will, however, be stated. Starting at transmitter input we have:

CRC processor: Receives transport blocks from the MAC layer. For transport blocks larger than 3824 bits, a 24-bit CRC is added. For transport blocks less than or equal to 3824 bits, a 16-bit CRC is used to reduce overhead.

Coding: LDPC coder. See Sect. 6.3.4.

Modulation symbol mapper:

- For the DL with CP-OFDM: QPSK, 16-QAM, 64-QAM, 256-QAM.
- For the UL with CP-OFDM: QPSK, 16-QAM, 64-QAM, 256-QAM.
- For the UL with DFTS-OFDM: $\pi/2$ BPSK, QPSK, 16-QAM, 64-QAM, 256-QAM. DFTS-OFDM is optional and used in link budget-limited cases.

Layer mapper:

- For the DL: One coded transport block mapped on up to four layers. If there is a second coded transport block, it is mapped on up to four more layers.
- For the UL: One coded transport block mapped on up to four layers.

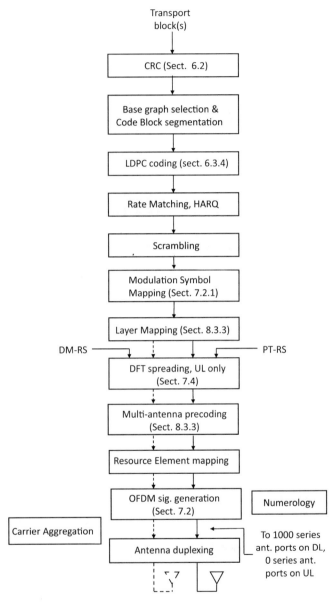

Fig. 10.9 PDSCH and PUSCH processing

OFDM signal generator:

– FFT size: 512 minimum, 4094 maximum.
– Subcarrier spacing: 15 kHz minimum, 120 kHz maximum.

Resource and physical antenna mapper: Broadly speaking, modulation symbols to be transmitted on each antenna port can be mapped to any resource element not specifically designated to other physical channels or signals. Resource elements in turn are mapped to the 1000 series antenna port(s) on the DL and 0 series antenna ports on the UL.

In the receiver, the processing is the inverse of that in the transmitter.

10.10.2   Physical Broadcast Channel (PBCH) Processing

Shown in Fig. 10.10 is a block diagram summary of the various steps taken by the PBCH on the transmit side to process transport blocks received from the BCH containing MIB data. The transport blocks are of size 32 bits and arrive at the PBCH processor at the rate of one every 80 ms. Note that there is no layer mapper in PBCH processing as PBCH uses a single antenna port transmission scheme. It uses the same antenna port as the PSS and the SSS. Starting at the physical layer input, we have:

CRC processor: A 24-bit CRC is added to each received transport block.

Coding: Polar coder. See Sect. 6.3.5.

Rate matcher: The output sequence length is 864 bits.

Modulation symbol mapping: QPSK modulation. The output stream is 432 complex modulation symbols.

Resource and physical antenna mapper: The PBCH together with the PSS and SSS are transmitted together in a block referred to as a *synchronization signal block* or SS block. The SS block is transmitted on a set of defined resource elements as shown in Fig. 10.11. The total number of resource elements used per SS block for

Fig. 10.10 PBCH transmitter processing

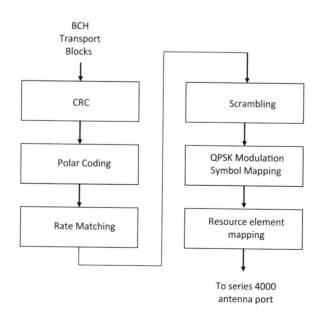

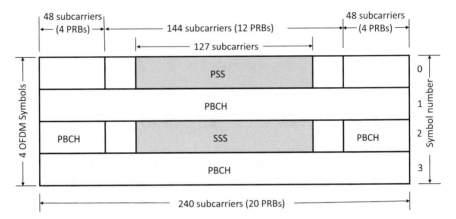

Fig. 10.11 Frequency/time structure of an SS block

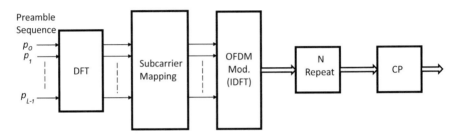

Fig. 10.12 Structure for RACH construction

PBCH transmission is 576 which includes 144 for the Demodulation Reference Signals (DM-RS) required for the coherent demodulation of the PBCH. Resource elements are mapped to the series 4000 antenna port.

Beam sweeping can be applied to SS block transmission. With such beam sweeping, SS blocks are transmitted in different beams in successive times, the set of such blocks within a beam sweep being referred to as an SS burst set.

In the receiver, the processing is the inverse of that in the transmitter as shown in Fig. 10.10.

10.10.3   Physical Random-Access Channel (PRACH) Processing

Shown in Fig. 10.12 is block diagram of the structure used to generate random-access preambles in NR. A preamble is generated by commencing with a length L Zadoff–Chu sequence, such a sequence consisting of complex symbols and exhibiting the property of constant power in both the frequency and time domain. With a prime length Zadoff–Chu sequence of length L, L-1 different sequences can be generated. As shown in Fig. 10.12, the sequence is DFT precoded, mapped to the designated subcarriers, and then applied to an OFDM modulator. Next, the modulator output is repeated N times after which a cyclic prefix is added.

NR specifies two types of preambles, namely, *long preambles* and *short preambles*, based on the length of the preamble sequence.

Long preambles have a sequence length $L = 839$ and a subcarrier spacing of either 1.25 kHz or 5 kHz. Such preambles are used only for operation in the lower frequency bands (FR1). There are four different formats where for each format there is a specified subcarrier spacing, number of repetitions, and cyclic prefix length. The number of repetitions varies between 1 and 4, and the cyclic prefix length varies between 15 and 680 μs.

Short preambles have a sequence length $L = 139$ and a subcarrier spacing of either 15 kHz or 30 kHz for operation in the lower frequency bands (FR1), and a subcarrier spacing of 60 and 120 kHz for operation in the higher frequency bands (FR2). Here, there are nine formats, and the number of repetitions varies between 1 and 12, and the cyclic prefix length varies between 7 and 66.7 μs.

The PRACH preamble configuration to use is provided to the UE in previously received system information.

10.11 Physical Layer Processing of Control Channels

In this section, we examine how control information is processed at the physical layer, namely, by the Physical Downlink Control Channel (PDCCH) and the Physical Uplink Control Channel (PUCCH). As with transport channels covered above, time/frequency resource mapping, which can be quite detailed, is only addressed here in the most general terms. For those interested in a more detailed description, it is well covered in [4].

10.11.1 *Physical DL Control Channel (PDCCH) Processing*

Shown in Fig. 10.13 is the physical layer processing by the PDCCH. The PDCCH carries *Downlink Control Information* (DCI) from the gNB to the UE. DCI includes information on where in the DL resource assignments have been scheduled; information to enable the UE to properly receive, demodulate, and decode the DL-SCH; and information to the UE about UL scheduling allotments and the specific resources and transport format to use for UL-SCH transmission, power control commands, etc. Starting at the input to the PDCCH processor, we have:

CRC processor: A 24-bit CRC is added to each received transport block. After attachment, the CRC parity bits are scrambled with the corresponding 16-bit *Radio Network Temporary Identifier* (RNTI). The RNTI can contain the identity of an

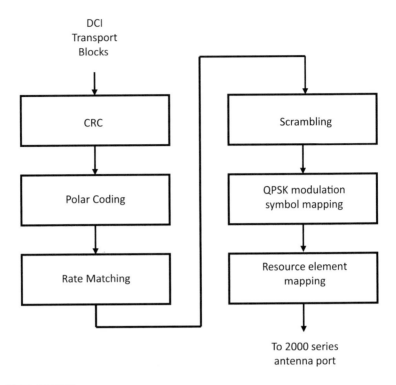

Fig. 10.13 PDCCH processor

intended device, identity of a group of UEs in the case of paging, identity of a group of UEs for which power control is issued, etc.

Coding: Polar coder. See Sect. 6.3.5.

Rate matcher: The rate matching is done per coded block and consists of sub-block interleaving, bit selection, and bit interleaving. It is done to set the number of coded bits to the resources available for PDCCH transmission. In sub-block interleaving, the incoming block is divided into 32 sub-blocks, and these sub-blocks interleaved. Bit selection is the true rate matching. It is achieved by feeding the output of the sub-block interleaver of length N bits into a circular buffer of length N and feeding data out with either no puncturing or puncturing, shortening, or repetition. This selection process is shown in Fig. 10.14. If E coded bits are required for transmission and $E = N$, then no rate matching is applied. If $E < N$, then puncturing or shortening is applied. Puncturing is achieved by selecting from the buffer bits from position $N - E$ to position $N - 1$, while shortening is achieved by selecting bits for transmission from position 0 to position $E - 1$. If $E > N$, then repetition is applied and is achieved by selecting all N bits from the buffer and repeating $E - N$ consecutive bits from the buffer.

Modulation Symbol Mapper: QPSK modulation.

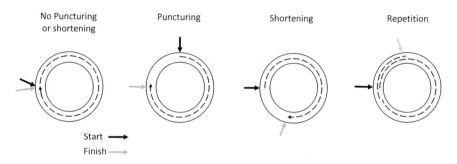

Fig. 10.14 PDCCH rate matching

Resource and physical antenna mapper: The time-frequency resources upon which the PDCCHs are transmitted are referred to as *Control Resource Sets* (CORSETs). In the frequency domain, a CORSET is a set of contiguous or discontinuous physical resource blocks (PRBs), structured in multiples of six PRBs, and can be configured anywhere in the frequency range of the carrier. In the time domain, each is of length one to three OFDM symbols and can be configured at any position within a slot. The PDCCH is transmitted on the 2000 series antenna port.

10.11.2 *Physical UL Control Channel (PUCCH) Processing*

The PUCCH carries *Uplink Control Information* (UCI) from the UE to the gNB. UCI information includes H-ARQ acknowledgments of received DL-SCH transport blocks, Channel State Information (CSI) regarding the DL channel status, and scheduling requests by the UE to be granted UL resources for UL-SCH transmission. It will be observed on Fig. 10.4 that the UCI is shown directed to both the PUCCH and the PUSCH. This is because NR allows the transmission of the UCI on the PUSCH if the device is transmitting data on the PUSCH. When this is the case, the UCI is multiplexed with the data on the allocated resources. If no data is being transmitted, then the UCI is transmitted on the PUCCH. The UE can beam-form the PUCCH by essentially transmitting it on the same beam as it is used for receiving the corresponding downlink transmission.

Five PUCCH formats are specified in NR. Two formats, Formats 0 and 2, are referred to as Short PUCCH formats. They occupy a maximum of two OFDM symbols that are transmitted in the last one or two symbols of a slot. Format 0 is capable of transmitting 2 or less bits, while Format 2 is capable of transmitting more than 2 bits. They support very fast feedback of HARQ acknowledgments. The other three formats, Formats 1, 3, and 4, are referred to as long PUCCH formats. They occupy from 4 to 14 OFDM symbols and are all capable of transmitting more than 2 bits. The long formats are used when better coverage than that afforded by the short formats is required as the long format results in higher received energy and hence more reliable reception. The following are structures of the various formats.

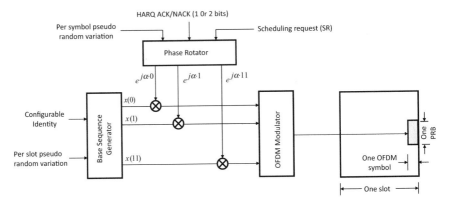

Fig. 10.15 PUCCH Format 0 structure

10.11.2.1 PUCCH Format 0 Structure

Shown in Fig. 10.15 is the PUCCH Format 0 structure. It is used for HARQ acknowledgments and scheduling requests (SRs). Information is transmitted by creating a length 12, low PAPR, Zadoff–Chu base sequence and then applying different linear phase rotations in the frequency domain to the sequence generator output via a phase rotator to create different sequences.

The base sequence used is configured per cell using an identity provided as part of the system information. In addition, it is varied randomly on a per slot basis to randomize interference between different cells.

As shown in Fig. 10.15, one input is HARQ ACK/NACK. If only one DL transport block was transmitting in a TTI (four layers or less), then only one ACK (block successfully decoded) or NACK (block not successfully decoded) needs to be transmitted, i.e., only 1 bit is required. In this case, the phase rotation parameter α is set to either 0 or π. If two DL transport blocks were transmitted, then one of four possible sets of information needs to be transmitted, namely, NACK/NACK, NACK/ACK, ACK/ACK, and ACK/NACK. Thus, 2 bits are required. Here α is set to either 0, $\pi/2$, π, or $3\pi/2$.

In the case of a simultaneous scheduling request, the phase rotation α is increased by $\pi/4$ for 1-bit acknowledgments. Thus, α is set to either $\pi/4$ or $5\pi/4$. For 2-bit acknowledgments, α is increased by $\pi/6$. Thus, α is set to either $\pi/6$, $3\pi/2$, $7\pi/6$, or $5\pi/3$.

Just as the base sequence can be varied randomly per slot, so too can the phase rotation be varied per symbol.

Also shown in Fig. 10.15 is the structure of the Format 0 PUCCH in the time/frequency resource. In the time domain, it is shown occupying the last symbol but can also occupy the last two symbols. If two symbols are used, the same information is transmitted on both symbols. In the frequency domain, it consists of 12 subcarriers and hence occupies one or two PRBs.

We note that the PUCCH Format 0 is not accompanied by a DM-RS.

10.11.2.2   PUCCH Format 1 Structure

PUCCH Format 1 is, to a certain extent, the long format version of Format 0. It can carry 2 bits using 4 to 14 OFDM symbols, each symbol occupying one resource block in the frequency domain. It is used for HARQ acknowledgments, scheduling requests (SRs), or both. The OFDM symbols are split sequentially between those used for control data and those used as reference signals to facilitate coherent detection. The split between those used for control and those used for reference is typically close to even. Shown in Fig. 10.16 [4] is the PUCCH Format 1 structure. If the UCI data is a single bit, it is BPSK modulated; if it is 2 bits, then it is QPSK modulated, resulting in both cases in a complex valued symbol, $d(0)$. Per modulated symbol out, as with Format 0, a cyclic shift is applied to create a pseudo-random variation. This cyclic shifted variation is multiplied by a 12-bit sequence, of the same type used in the Format 0 creation, resulting in a length 12 sequence of complex valued symbols. Also, as with Format 0, this sequence is configured per cell using an identity provided as part of the system information and varied randomly on a per slot basis to randomize interference between different cells. The complex valued sequence is block-wise spread in the time domain with an orthogonal sequence of length equal to the number of OFDM symbols used for control information. In the example shown in Fig. 10.16, four symbols are used for control information; hence, a length 4 orthogonal sequence is used. By using different orthogonal codes, multiple UEs having the same base sequence and phase rotation and using the same

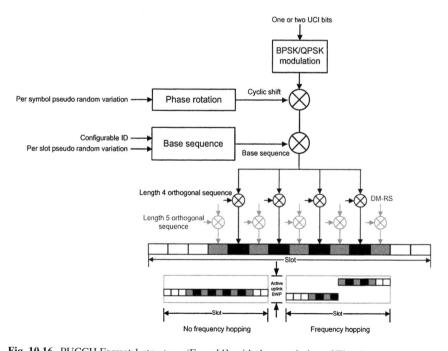

Fig. 10.16 PUCCH Format 1 structure. (From [4], with the permission of Elsevier)

resource can still be differentiated at the gNB, thus increasing the multiplexing capacity. For a PUCCH transmission spanning multiple slots, the complex valued symbol $d(0)$ is repeatedly used in creating the length 12 sequences for those slots.

PUCCH Format 1 can use frequency hopping to achieve a certain degree of frequency diversity, an example of such frequency hopping being shown in Fig. 10.16. The use of frequency hopping is configurable and determined by the gNB as part of the PUCCH resource configuration.

10.11.2.3   PUCCH Format 2 Structure

As indicated above, Format 2 is a short format PUCCH. It occupies a maximum of two OFDM symbols that are transmitted in the last one or two symbols of a slot. It is capable of transmitting more than 2 bits, carrying, for example, simultaneous HARQ acknowledgments and CSI reports, as well as a SR. If the payload is larger than can be accommodated, then HARQ acknowledgments are given priority over CSI reports. Shown in Fig. 10.17 is the PUCCH Format 2 structure. Let the payload size be N_{UCI}. Then, starting at the UCI input point, we have:

CRC processor: If $N_{UCI} \leq 11$, no CRC bits are attached. If $12 \leq N_{UCI} \leq 19$, 6 CRC bits are attached. If $N_{UCI} \geq 20$, 11 CRC bits are attached.

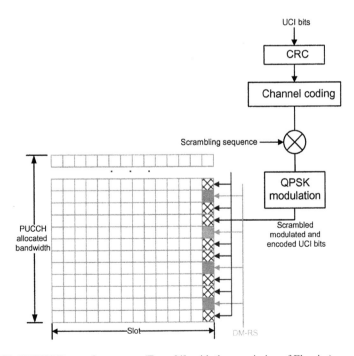

Fig. 10.17   PUCCH Format 2 structure. (From [4], with the permission of Elsevier)

Channel coder: $N_{UCI} \geq 12$, the UCI bits with CRC attachment are encoded with polar coding (Sect. 6.3.5). If $3 \leq N_{UCI} \leq 11$, then the UCI bits are encoded with Reed–Muller coding (Sect. 6.3.6.3). If $N_{UCI} = 2$, simplex coding is used (Sect. 6.3.6.2). If $N_{UCI} = 1$, repetition coding is used (Sect. 6.3.6.1).

Scrambler: The scrambling sequence is a function of both the UE identity and the physical layer cell identity.

Modulation symbol mapping: QPSK

Resource element mapping: The modulated symbols are sequentially mapped to subcarriers reserved for PUCCH transmission and not used by the associated DM-RSs. They are mapped across multiple resource blocks utilizing one or two OFDM symbols. The associated DM-RSs are mapped to every third subcarrier and consist of pseudo-random generated QPSK symbols. PUCCH Format 2 is normally transmitted at the end of a slot as shown in Fig. 10.17 [4], but depending on certain conditions, they may be transmitted in other positions within a slot.

10.11.2.4 PUCCH Format 3 Structure

PUCCH Format 3 is the long version of PUCCH Format 2. Like Format 2, it can transmit more than 2 bits, but here, these can be transmitted over 4 to 14 symbols using multiple resource blocks per symbol. The result is that this is the format with the largest payload capacity. Just as with Format 1, OFDM symbols used are split between those used for control data and those used as reference signals to facilitate coherent detection. Shown in Fig. 10.18 is the PUCCH Format 3 structure. As before, let the payload size be N_{UCI}. Then, starting at the UCI input point, we have:

CRC processor: Same as for Format 2.
Channel coder: Same as for Format 2.
Scrambler: Same as for Format 2.

Modulation symbol mapping: QPSK, with option to use $\pi/2$ BPSK.

Resource element mapping: Modulated symbols are divided between the OFDM symbols with $12 \times M_{RB}$ modulated symbols directed to each OFDM symbol in M_{RB} resource blocks, where M_{RB} denotes the number of resource blocks occupied by the PUCCH. These sets of modulated symbols are DFT spread to lower the resulting PAPR prior to OFDM modulation which results in its placement in the resource grid. Operation is configurable with frequency hopping as shown in Fig. 10.18 [4] to exploit frequency diversity, but also configurable without. Reference signal symbol placement is based on whether or not frequency hopping is used and also on the length of the PUCCH transmission as at least one reference signal must be utilized per hop.

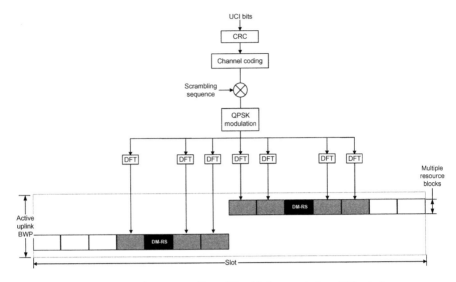

Fig. 10.18 PUCCH Format 3 structure. (From [4], with the permission of Elsevier)

10.11.2.5 PUCCH Format 4 Structure

PUCCH Format 4 is identical to Format 3 up to the point of modulation output. The difference, as shown in Fig. 10.19 [4], is in the modulated symbol mapping. Here, the modulated symbols are divided between (a) the OFDM symbols with $12/N_{SF}$ modulated symbols directed to each OFDM symbol, (b) each $12/N_{SF}$ set spread by a length N_{SF} block spreading sequence ($N_{SF} = 2$ or 4) to create a length 12 modulated symbol output, and (c) the length 12 symbol output DFT spread and OFDM modulated resulting in its placement in the resource grid. Frequency hopping and reference signal placement are similar to Format 3.

10.12 Physical Signal Processing

Physical signal processing is now addressed. Such signals are predefined, contain no information from higher layers, and occupy specified resource elements in the physical resource grid. NR physical signals are generated via three types of sequences, namely:

– *Maximum length* (m) sequences, which are pseudo-random binary sequences. They are generated using a shift register with linear feedback. A length n register produces sequences of length $2^n - 1$. It is possible to select a small subset of sequences created by the same shift register that have relatively low cross correlation.

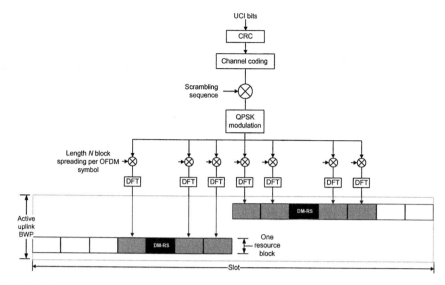

Fig. 10.19 PUCCH Format 4 structure. (From [4], with the permission of Elsevier)

- *Gold sequences*, which are also a type of pseudo-random binary sequences, constructed by adding together in a defined fashion two m sequences of the same length. A set of Gold sequences consists of $2^n - 1$ sequences, each one of length $2^n - 1$. Every sequence within a set has a low cross correlation with the other sequences in the set, making it easy in decoding to distinguish between the individual sequences even when corrupted by noise. If properly constructed, gold sequences have better cross-correlation properties than m sequences.
- *Zadoff–Chu sequences* (already mentioned above), which are complex value sequences that when used to linearly modulate a signal give rise to a signal of constant amplitude. Further, the DFT of such a sequence is also a Zadoff–Chu sequence. Such sequences exhibit the very useful property that cyclically shifted versions of themselves appear as orthogonal to one another at a receiver, given that each cyclic shift, as seen within the time domain of the signal, is greater than the propagation delay plus multipath delay spread of the signal between the transmitter and receiver.

As indicated above, NR physical signals are Demodulation Reference Signals (DM-RSs), Phase-Tracking Reference signals (PT-RSs), Channel State Information Reference Signals (CSI-RSs), and synchronization signals (SSs). Following is a description of the processing applied to create physical signals. Time/frequency resource mapping, which can be quite detailed, is only addressed here in the most general terms. For those interested in more detailed descriptions, they are well covered in [2, 4]. First DL physical signals are described followed by UL signals.

10.12.1   Downlink Physical Signals

10.12.1.1   Demodulation Reference Signal (DM-RS) for PDSCH

The DM-RS for PDSCH is transmitted together with the PDSCH on Series 1000 antenna ports and subject to the same precoding as the PDSCH. Its point of injection in the PDSCH processing is shown in Fig. 10.9. It is created via the generation of pseudo-random sequences. The generation process is as shown conceptually in Fig. 10.20. First, two slightly different length $2^{31}-1$ Gold sequences are created. Both unipolar sequences (0s and 1s) are then converted to bipolar format (-1s and +1s). One of the sequences is then rotated through 90^0 so as to be in quadrature with the other, and both sequences are then summed to create complex QPSK modulation symbols. As with PDSCH modulation symbols, these symbols undergo multi-antenna precoding, resource element mapping, OFDM signal generation, and antenna duplexing. The QPSK sequence is generated across all the resource blocks but transmitted only in the resource blocks used for data transmission as there is no need for having knowledge of the channel outside of the frequency spectrum occupied by the channel.

In NR, to help achieve low latency, a front-loaded structure is used whereby the DM-RSs are located early in the transmission. There are two main mapping structures in the time domain, namely, mapping type A and mapping type B. With type A, the first DM-RS is located in symbol 2 or 3 of the slot and is used in cases where the data uses most of the slot. With type B, the first DM-RS is located in the first symbol where data is allocated. To assist high-speed situations, up to three additional DM-RSs can be configured in a slot for both types A and B.

Two different types of DM-RSs can be configured, type 1 and type 2. With both types, the underlying pseudo-random sequence is mapped to every second subcarrier in the OFDM symbol used for DM-RS transmission. To support multilayer MIMO transmission, multiple orthogonal signals, each directed to a different series 1000 antenna port, can be created by multiplying the underlying sequence with different length 2 orthogonal sequences in the frequency domain. With type 1, up to four orthogonal reference signals can be provided with a single-symbol DM-RS and up to eight with a double-symbol DM-RS. With type 2, up to 6 orthogonal reference signals can be provided with a single-symbol DM-RS and up to 12 with a double-symbol DM-RS, thus supporting a higher degree of multiuser MIMO than with type 1.

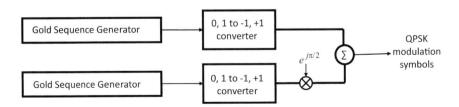

Fig. 10.20   PDSCH-associated DM-RS generation

10.12.1.2   DM-RS for PDCCH

The DM-RS for the PDCCH is transmitted together with the PDCCH. It is generated in a similar fashion to the DM-RS for PDSCH where a pair of different length $2^{31}-1$ Gold sequences is used to derive a QPSK modulated symbol stream. The QPSK sequence is generated across all the resource blocks but transmitted only in the resource blocks used for PDCCH transmission, being mapped onto every fourth subcarrier in an occupied resource group, and directed to the single antenna port used for PDCCH transmission, namely, antenna port 2000.

10.12.1.3   DM-RS for PBCH

The DM-RS for the PBCH is transmitted together with the PBCH. It is generated in a similar fashion to the DM-RS for PDSCH where a pair of different length $2^{31}-1$ Gold sequences is used to derive a QPSK modulated symbol stream. The QPSK sequence is generated across all the resource blocks but transmitted only in the resource blocks used for PBCH transmission, being mapped onto every fourth subcarrier in an occupied resource group, and directed to the single antenna port used for PBCH transmission, namely, antenna port 4000.

10.12.1.4   Downlink Phase-Tracking Reference Signal (PT-RS)

The DL Phase-Tracking Reference Signal (PT-RS), like the DL DM-RS for the PDSCH, is transmitted together with the PDSCH, its point of injection in the PDSCH processing being as shown in Fig. 10.9. It is present only if it is explicitly configured. It may not be configured, for example, in low-frequency transmission where common phase error may be negligible. It is generated in a similar fashion to the DM-RS for PDSCH where a pair of different length $2^{31}-1$ Gold sequences is used to derive a QPSK modulated symbol stream. The QPSK sequence is generated across all the resource blocks but transmitted only in the resource blocks used for PDSCH transmission. Its mapping is designed to have low density in the frequency domain and high density in the time domain. This is because the phase rotation caused by common phase error is the same for all subcarriers in an OFDM symbol, whereas it has low correlation across OFDM symbols.

In the frequency domain, the PT-RS is transmitted in every second or fourth resource block. In the time domain, the first PT-RS is repeated every first, second, or fourth symbol starting with the first PT-RS symbol in the allocation.

The antenna port used for PT-RS transmission is the lowest numbered 1000 series port used in the DM-RS antenna port group.

10.12.1.5   Channel State Information Reference Signal (CSI-RS)

The CSI-RS is generated in a similar fashion to the DM-RS for PDSCH where a pair of different length $2^{31}-1$ Gold sequences is used to derive a QPSK modulation symbol stream. The QPSK sequence is generated across all the resource blocks but transmitted only in the resource blocks used for PDSCH transmission.

A configured CSI-RS can correspond to up to 32 antenna ports and is transmitted on the Series 3000 antenna ports. In the time domain, it may commence at any OFDM symbol of a slot and extend over 1, 2, or 4 OFDM symbols depending on the number of antenna ports configured. It is always configured on a per UE basis. However, this does not necessarily mean that a configured CSI-RS can only be used by a single UE, but rather that the same set of CSI-RS resources can be configured separately for several devices. A single-port CSI-RS occupies a single resource element within one resource block in the frequency domain and one slot in the time domain and can be configured to occupy any position within the resource block that doesn't collide with other DL physical channels and signals.

A multi-port CSI-RS can be regarded as multiple CSI-RSs, orthogonal to each other per antenna port, sharing the same set of resource elements specified for transmission of the configured multi-port CSI-RS. This resource sharing is accomplished, in general, via a combination of:

- Code-domain sharing, where different per antenna port CSI-RS are transmitted using the same set of resource elements, differentiation achieved by spreading each CSI-RS with different orthogonal codes
- Frequency-domain sharing, where different antenna port CSI-RS are transmitted on different subcarriers over an OFDM symbol
- Time-domain sharing, where different antenna port CSI-RS are transmitted on different OFDM symbols within a slot

Unlike the DM-RS, the CSI-RS does not undergo multi-antenna precoding that's applied to user data. Rather, the CSI-RS ports are mapped directly to either the gNBs physical antennas or via a spatial filter that maps M CSI-RS ports to N physical antennas. When such a filter is used, it is seen by the UE as an integrated part of the overall channel, and thus the channel being sounded is not the actual, physical channel. The mapping via the spatial filter allows different CSI-RSs to be beamformed in different directions.

10.12.1.6   Tracking Reference Signal (TRS)

The TRS is not a CSI-RS per say. Rather, it is a resource set consisting of multiple periodic NZP CSI-RSs and transmitted on the Series 3000 antenna ports. As shown in Fig. 10.21 [2], the resource set is configured within one resource block in the frequency domain and two consecutive slots in the time domain. The set consists of four single-port CSI-RSs, each occupying three resource elements in an OFDM symbol, with the separation between two CSI-RSs within a slot always being four

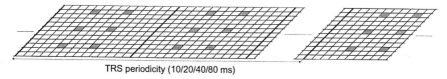

TRS periodicity (10/20/40/80 ms)

Fig. 10.21 TRS resource plane structure. (From [2], with the permission of Elsevier)

symbols. The set can be configured with a periodicity of 10, 20, 40, or 80 ms. An alternative structure is defined with the same per slot structure as shown Fig. 10.21 but with the set consisting of two CSI-RSs in only one slot.

10.12.1.7 Primary Synchronization Signal (PSS)

The PSS is generated from a length 127 m sequence shift register. By applying different shifts to the basic m-sequence, three different PSSs are generated. The 0s and 1s outputted from the sequence generating shift register are used to create BPSK modulation symbols of value -1 and +1. These symbols are fed to the OFDM processor which resource maps them to 127 subcarriers as part of an SS block and as shown in Fig. 10.11. They are mapped to antenna port 4000 along with the PBCH.

10.12.1.8 Secondary Synchronization Signal (SSS)

The SSS is generated from a length 127 Gold sequence generator. A total of 336 different sequences are defined. The 0s and 1s outputted from the sequence generator are used to create BPSK modulation symbols of value -1 and +1. As with the PSS, these symbols are fed to the OFDM processor which resource maps them to 127 subcarriers as part of an SS block and as shown in Fig. 10.11. They are mapped to antenna port 4000 along with the PBCH.

As the PSS can be one of three different sequences and the SSS one of 336 different sequences, together they are able to convey 1008 (3×336) unique physical layer cell identities.

10.12.2 Uplink Physical Signals

10.12.2.1 DM-RS for CP-OFDM PUSCH

In NR, the same DM-RS as used in the DL, i.e., Gold sequence-derived QPSK modulation symbols, is used in the UL for the CP-OFDM case. Thus, the description given in Sect. 10.12.1.1 above is equally applicable here if we substitute PUSCH for PDCSH and antenna port series starting with 0 for that starting with 1000.

10.12.2.2   DM-RS for DFTS-OFDM PUSCH

DFTS-OFDM is configured for single-layer transmission only and is designed to be used primarily in situations where coverage is challenging. Here, the DM-RS is based on Zadoff–Chu sequences and supports continuous allocations. Multiple sequences are generated from a single-base sequence by applying different linear phase shifts in the frequency domain as shown in Fig. 10.22. Defined OFDM symbols within a slot are assigned exclusively for DM-RS transmission, following the same mapping as configuration of type 1 defined in Sect. 10.12.1.1 for the DL DM-RS. Type 2 mapping is not supported as there is no requirement for handling a high degree of multi-used MIMO.

10.12.2.3   DM-RS for PUCCH Format 1

The DM-RS for PUCCH Format 1 is created from a length 12 unmodulated Zadoff–Chu sequence. It is inserted in the time domain as shown in Fig. 10.16, the sequence being block-wise spread in the time domain with an orthogonal sequence of length equal to the number of OFDM symbols used for reference signals. In Fig. 10.16, five OFDM symbols are used for reference signals, hence the length 5 orthogonal sequence used for block-wise spreading. Resource mapping is to antenna port 2000 series.

10.12.2.4   DM-RS for PUCCH Format 2

The DM-RS for PUCCH Format 2 has the same structure as that used for the PDSCH and is thus gold sequence-derived QPSK modulation symbols. These symbols are mapped to every third subcarrier in each OFDM symbol as shown in Fig. 10.17. Resource mapping is to antenna port 2000 series.

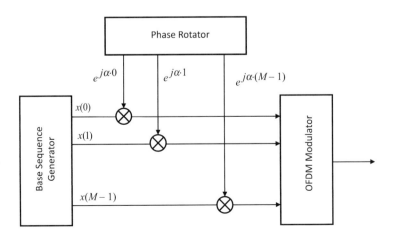

Fig. 10.22 Generation of DFTS-OFDM-associated UL DM-RS

10.12.2.5   DM-RS for PUCCH Formats 3 and 4

The DM-RS for PUCCH Formats 3 and 4 is generated in the same way as for DFTS-OFDM PUSCH transmissions described in Sect. 10.12.2.2. The placement of reference signal symbols depends on whether or not frequency hopping is used as well as the length of the PUCCH transmission given that there must be at least one reference signal per frequency hop. Resource mapping is to antenna port 2000 series.

10.12.2.6   Uplink PT-RS for CP-OFDM

In NR, the same PT-RS structure as used in the DL is used in the UL for the CP-OFDM case. Thus, the description given in Sect. 10.12.1.4 above is equally applicable here if we substitute PUSCH for PDCSH and antenna port series starting with 0 for that starting with 1000.

10.12.2.7   Uplink PT-RS for DFTS-OFDM

The UL PT-RS for DFTS-OFDM is generated via Gold sequences to create a QPSK modulation symbol stream. These symbols are inserted prior to DFT precoding, and the time domain mapping is the same as the CP-OFDM case.

10.12.2.8   Sounding Reference Signal (SRS)

The SRS is configured to allow the gNB to estimate the UL channel in much the same way that the CSI-RS is configured to allow the UE to estimate the DL channel. The SRS is designed to have low PAPR, this being achieved by employing sequences partly based on Zadoff–Chu sequences. It can, in general, cover one, two, or four consecutive OFDM symbols, being located within the last six symbols of a slot. In the frequency domain, it has a comb-like structure, where it is transmitted on every second or fourth subcarrier, this being referred to as comb-2 or comb-4, respectively. Transmission from different UEs can take place within the same frequency range by each using a comb pattern that corresponds to different subcarriers than those used by the rest. With comb-2, two SRSs can be so accommodated, while for comb-4, up to four SRSs can be accommodated.

When an SRS supports more than one antenna port, the different ports utilize the same basic SRS sequence and the same set of resource elements. However, the signals transmitted from different ports are differentiated from each other by applying different phase rotations to the symbols on the different ports.

Up to four Series 1000 antenna ports are supported, being configurable to either 1, 2, or 4 ports. Unlike the DM-RS, the SRS does not undergo multi-antenna precoding that's applied to user data. Rather, similar to CSI-RS, the SRS ports are either mapped directly to the UEs physical antennas or via some spatial filter that maps M SRS ports to N physical antennas. This mapping via the spatial filter allows beamforming of the transmitted signal. When such a filter is used, it is seen by the gNB as an integrated part of the overall channel.

10.13   Initial Access

Initial access are the steps executed to allow a UE that is just powered on to find a cell to camp on and, having found such a cell, the steps that the UE, in the idle or inactive RRC state, uses to access the network and, via a random-access procedure, establishes normal communication, i.e., go to the RRC-CONNECTED state [2]. A random-access procedure can be contention-based or contention-free, the latter used only if the UE is already known to the network and has been allocated pre-ambles. Shown in Fig. 10.23 and described below is a simplified version of the initial access steps where the random access is contention-based:

1. The gNB periodically transmits SS blocks on the PBCH and SIBs on the PDSCH.
2. UE performs initial cell selection and DL synchronization. Initial cell selection involves finding a strong PBCH signal (if the PBCH is being beam swept, then this means finding the strongest beam). It does this by scanning all RF channels within its capability. On each carrier frequency, it searches for the strongest cell. Once a suitable cell is found, this cell is selected. It uses the PBCH signal of this cell for establishing a connection, obtaining an estimate of frame timing, obtaining cell identification, and finding the PSS and SSS necessary for the coherent demodulation of the PCBH and PDCCH. First the UE tries to find and demodulate the PSS so as to obtain symbol and half-frame timing. If successful, it then

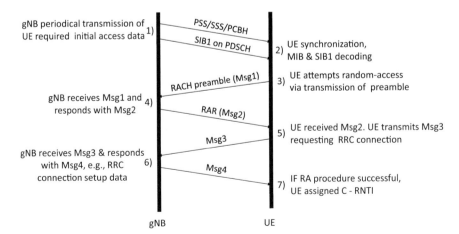

Fig. 10.23 Initial access procedure

attempts to demodulate the SSS so as to determine the cyclic prefix length, the duplexing scheme, and the exact frame timing. It then detects the physical cell identity from the sequences used on the PSS and SSS and decodes the PBCH to obtain the MIB. With the aid of the MIB, it is able to receive the PDSCH and PDCCH. It decodes the PDSCH and obtains SIB1 which allows it to commence access to the system.

3. UE attempts random access by transmitting a preamble (Msg1) on the PRACH channel. If no response is received from the network within a defined window, the UE assumes that the preamble was not correctly received and resends it at a higher transmit power level and continues this process until a response is received or a maximum number of tries is reached.

4. The gNB receives Msg1 and responds with Msg2, a Random-Access Response (RAR) via the PDCCH/PDSCH. Included in the response is a scheduling grant indicating resources the UE can use for the transmission of its next message, Msg3, a timing correction determined by the network based on the timing of the received preamble (Sect. 10.14), and a temporary identity, the Temporary Cell Radio Network Temporary Identifier (TC-RNTI). Had the network received random-access attempts from several devices, the individual response messages are combined in a single transmission. Had all the devices used different preambles, then resources allocated for upcoming uplink transmission would be different, and no collision would occur. However, should multiple devices have used the same preamble, then a collision or collisions would occur. The following steps resolve this collision(s) dilemma.

5. The UE receives Msg2 and adjusts its uplink transmission timing. For the UE to be able to transmit user data, i.e., become RRC-CONNECTED, it needs to be assigned a unique identity within the cell, the C-RNTI. The UE transmits the necessary information to make this possible, including a UE identity, over the UL-SCH, Msg3, in the resources assigned in the DL Msg2.

6. The gNB receives Msg3 and responds with a contention resolution message, Msg4, addressed using the TC-RNTI and intended to ensure that a UE does not incorporate another UE identity and, if so, become RRC-CONNECTED.

7. The UE compares the identity in the message with the identity transmitted in Msg3. If these identities match, the random-access procedure is declared successful, the TC-RNTI is redefined as the C-RNTI, and the UE becomes RRC-CONNECTED.

10.14 Uplink Timing Correction

As indicated in Sect. 10.13 above, the gNB, during the random-access procedure, sends a timing correction to the UE determined by the gNB based on the timing of the received preamble. This timing feedback by the gNB is then maintained during the connection as frequently as needed. The purpose of timing alignment is to assure as best as possible that the uplink slot boundaries for a given numerology from

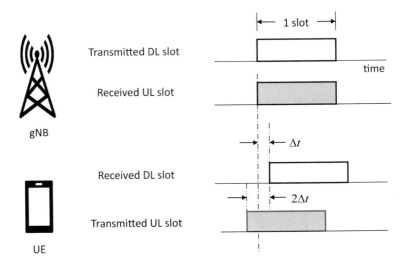

Fig. 10.24 Uplink timing advance

different UEs align at the gNB in order to guarantee uplink orthogonality. This is assured if any offset after adjustment falls within the cyclic prefix. Following initial access, the gNB can use the sounding reference signals to estimate the required offset but in principle can use any signal received from the UE. If a timing correction is required, the gNB transmits to the UE a timing advance command as a MAC control element on the DL-SCH.

As shown in Fig. 10.24, timing advance is a negative time offset, at the UE, to the start of an uplink slot relative to the start of the most recently downlink slot as observed by the UE. As shown in the figure, the downlink slot, when received by the UE, suffers a propagation delay of Δt relative to its start time at the gNB. Given that the uplink slot will suffer the same propagation delay, then, in order for it to arrive at the gNB time aligned, it needs to be offset relative to the received time of the downlink slot by $2\Delta t$. Clearly timing advance is a function of propagation delay and hence the distance between the gNB and the UE. The timing advance is conveyed in incremental steps, the individual step being a function of the subcarrier spacing. The larger the subcarrier spacing, the shorter the slot, and the smaller the timing advance step.

10.15   Uplink Power Control

The primary purpose of uplink power control is to minimize interference to cells other than the UE's cell. Interference within the UE's cell is a lesser issue given that transmissions within the same cell are ideally orthogonal. It also serves the purpose of minimizing UE power consumption. Uplink power control is the procedure whereby the transmit power at the UE of different physical channels and signals is adjusted so as to be received by its target gNB at a level necessary, but not excessively above, for proper decoding while at the same time being such that they are received throughout the surrounding cells at levels that don't result in undue interference. By a received level necessary for proper decoding, we mean a level resulting in the signal to noise necessary to meet the required BER. This level is a function of modulation complexity, higher complexity requiring a higher signal-to-noise ratio for the same BER. In achieving the purposes stated above, power control in a mobile environment has to continually react to the varying characteristics of the propagation channel, including distance-induced path loss, shadowing, fast fading, and interference from other users both within the cell and neighboring cells.

In NR, there are two types of uplink power control operations, open-loop and closed-loop:

- Open-loop power control, where the power is set by the UE based on the downlink received signal strength. This form of control provides an initial power setting for the UE prior to it establishing an RRC connection with the gNB, after which closed-loop operation can commence.
- Closed-loop power control, where the power is set based on explicit signaling from the gNB. The power commands contained in the signaling are based on the gNB measurements of the received uplink power, hence the "closed-loop" nomenclature. This form of power control provides finer granularity and is normally more effective than open-loop control, particularly given that it is based on actual measured uplink loss, not an estimate based on the assumption that uplink loss is similar to the downlink loss, an assumption that may not necessarily be correct in the case of FDD.

10.16 Low Latency

A key feature of NR is its ability to be configured for low latency [2, 3]. This feature is of particular importance in ultra-reliable low-latency communication (URLLC) service, an NR service category to support latency-sensitive services such as remote control and autonomous driving. In 3GPP TR 38.913, user plane latency is defined as "the time to successfully deliver an application layer packet/message from the radio protocol layer 2/3 SDU ingress point to the radio protocol layer 2/3 SDU egress point via the radio interface in both uplink and downlink directions, where neither device nor the Base Station reception is restricted by DRX." In simpler language, this means the time from when IP packets enter the RAN to when they leave, in the UL and DL directions, given no discontinuous reception. Physical layer latency results from the addition of many contributing components:

- The processing time at the transmit end to get ready for transmission, including encoding, etc.
- Average buffering time in the transmission buffer until the next TTI starts. This time is half the TTI time

Table 10.3 NR DL and UL
user plane latency

Delay component	Downlink latency (ms)	Uplink latency (ms)
TTI	0.125	0.125
Frame alignment	0.063	0.063
Transmit side processing	0.160	0.320
Receive side processing	0.170	0.090
Total time	**0.518**	**0.598**

– The TTI, the time to transmit a packet
– The signal propagation time from transmitter to receiver
– The processing time at the receive end to decode received data
– Retransmission delay, if any

For eMBB, the user plane latency target is 4 ms for UL and 4 ms for DL. For URLLC, the target is 0.5 ms for UL and 0.5 ms for DL. As way of comparison, the lowest equivalent latency for 4G LTE is approximately 5 ms. Thus, NR seeks to improve latency by a factor of about 10 in the extreme case.

The primary way in which NR addresses the low-latency requirement is, as was described in Sect. 10.7, to allow a minimum standard slot length and hence TTI of 0.125 ms. Further, by the use of mini-slots, this number can be reduced further. For 4G LTE, the TTI is 1 ms. The shorter minimum TTI in NR means the transmission time per packet is shorter which in turn reduces the buffering time in the transmission buffer and shortens the processing time. Shorter processing time is further aided by the use of "front-loaded" reference signals as was described in Sect. 10.12 and control signaling that conveys scheduling information at the beginning of the slot, allowing the receiving device to start processing the received data without prior buffering. Using data from [3], Table 10.3 has been constructed to show DL and UL NR RAN user plane latency, with the UE configured for low-latency services, and 120 kHz subcarrier spacing and hence a TTI of 0.125 ms. Note that the signal propagation time from transmitter to receiver, which is a function of transmission distance, is not shown in the table. For a 100-meter path, this propagation time is only 0.33 μs, and for a 10 km path, it is 0.033 ms.

10.17   Scheduling

Scheduling is key function of NR (and any multiuser mobile system) and to a significant extent determines the overall behavior of the system. Stated in simple terms, the scheduler determines, for each TTI, to which users the shared time-frequency resource is to be assigned and what data rate to apply in transmission. Both uplink and downlink transmissions are scheduled, and thus in the gNB, there is both a DL and UL scheduler. The scheduler is a part of the MAC layer, controlling the

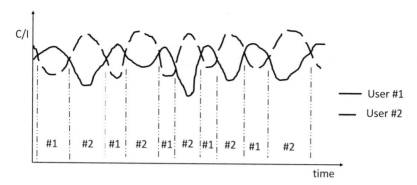

Fig. 10.25 Max C/I allocation

assignment of transmission resources in the form of resource blocks in the frequency domain and OFDM symbols in the time domain.

NR scheduling design is not specified by 3GPP, leaving this up to the system provider. The goal of such design is to factor in the instantaneous channel variations experienced by the various user channels in both the downlink and uplink directions and schedule transmissions that take advantage of these conditions in such a way as to assign resources to the channels that can best use them, such scheduling being referred to as *channel-dependent scheduling.*

Before looking, at a high level, as to how scheduling is implemented in NR, a review of three fundamental scheduling schemes is in order. These such schemes are round-robin (RR), maximum Carrier-to-interference ratio (Max C/I), and proportional fair (PF).

With *round-robin* (RR), radio resources are sequentially allocated among users. It is thus very fair but at the expense of overall system throughput as it may allocate resources to a user or users whose channel conditions do not permit effective data transfer. It most certainly is not channel-dependent scheduling.

With *maximum carrier-to-interference ratio* (Max C/I), users with the highest C/I and hence highest instantaneous achievable data rate are scheduled during the current scheduling decision interval. This results in the highest system throughput but does not provide any kind of fairness among users. Figure 10.25 shows a case of two users experiencing differing C/I and the time allocation to these users under the Max C/I scheme.

Proportional fair (PF) provides a good trade-off between RR and Max C/I, and its features, or slight variations thereof, are thus often implemented. Here users are scheduled according to the ratio of their C/I to their average served data rate. Thus, it schedules not only users with the best channel conditions but also those experiencing a low average data rate because of their channel conditions. This results in all users having a relatively equal probability of being served even though they experience very different instantaneous throughput rates. Those users experiencing a high C/I will be scheduled with a high coding rate and modulation complexity resulting in a high-throughput rate, whereas those experiencing a low C/I will be scheduled with a low coding rate and modulation complexity resulting in a low, but reliable, throughput rate. In addition to C/I and average served data rate, other factors, however, may need to be considered in scheduling such as quality of service guarantees granted to various users.

In NR, downlink and uplink scheduling are separate functions, and their scheduling decisions can be taken independently of each other.

The downlink scheduler dynamically controls the UEs to be transmitted to. Each scheduled UE is informed via the PDCCH of the time-frequency resources upon which the UE's DL-SCH is to be transmitted and the associated transport format, i.e., transport block size, the modulation and coding scheme, and antenna mapping. Downlink scheduling relies on knowledge of the downlink channel obtained via the downlink CSI-RS and reported back up to the gNB.

The uplink scheduler is similar to the downlink one in that it too dynamically controls which UEs are permitted to transmit on their UL-SCH. Each scheduled UE is provided by the PDCCH with a scheduling grant informing it of the time-frequency resources upon which the UE's UL-SCH is to be transmitted and the associated transport format. Uplink scheduling relies on knowledge of the uplink channel obtained from the uplink SRS. There is one difference, however. Here, the gNB does not explicitly schedule a given logical channel, but only schedules broadly the device, leaving it to the device to select from which radio bearers data will be transferred using a defined set of rules, designed to ensure that the UE satisfies the QoS of each radio bearer in an optimal way.

Though dynamic scheduling is the normal mode of scheduling, in certain cases, it is possible to configure transmission without a dynamic grant, in order to reduce control signal overhead.

In the downlink, semi-persistent scheduling is supported. Here, a semi-static scheduling pattern is communicated in advance to the UE via RRC signaling. A downlink assignment is provided by the PDCCH indicating required information such as time-frequency resources and transport format and activated/deactivated via Layer 1 signaling. Upon activation, the UE receives downlink data transmission according to the RRC communicated scheduling pattern.

In the uplink, two types of transmission without a dynamic grant are supported, namely:

– Grant type 1, where the grant, including all necessary transmission parameters, is provided by the RRC which also activates/deactivates the uplink transmission.
– Grant type 2, where the periodicity of transmission is provided by the RRC, but Layer 1/Layer 2 control signaling is used to activate/deactivate the transmission. As with downlink persistent scheduling, receiving the activation command the UE transmits according to the periodicity provided by the RRC if there is data in the buffer.

Both schemes reduce control signaling overhead as well as latency given that once a UE has data to send, it can immediately commence transmission as there is now no need to first send a scheduling request and await a grant.

10.18   Spectrum for 5G

5G NR is conceived to operate ultimately in spectrum ranging from approximately 400 MHz up to about 90 GHz. It is being designed to be able to operate in licensed, unlicensed, and shared frequency bands, shared bands being those where 5G shares the spectrum with nonmobile service providers. It operates in both the FDD and TDD modes.

3GPP defines two frequency ranges for NR, namely, FR1 and FR2. FR1 was specified in 3GPP TS 38.104 V15.1.0 to cover the range 410 to 6000 MHz. However, in V15.5.0, the range was extended covering 410 to 7125 MHz to allow it to be used in unlicensed bands in the 6 GHz region. FR2 was specified in 3GPP 38.104 V15.1.0 to cover the range 24.25 to 52.6 GHz, frequency bands in this range being referred to as millimeter wave (mmWave) bands. Spectrum wise, the FR2 range is the distinguishing feature of 5G relative to 4G. It allows for transmission in bands with much more spectrum than available in FR1 enabling high capacity and data rates.

The lowest frequency bands currently specified in FR1 is for FDD operation, covering approximately 600–800 MHz. Total spectrum in these bands is on the order of 15 to 40 MHz per direction and thus, as we shall see below, result in much lower capacity relative to FR2 bands. The big advantage of such bands, however, is wide area coverage, typically on the order of tens of kilometers, and low outdoor-to-indoor penetration loss.

In the 1.4–2.7 GHz region of FR1, the operation is mostly FDD with typical maximum spectrum per operating frequency being 20 MHz per direction. With carrier aggregation, up to 16 such carriers can be aggregated. Such a large number of aggregated carriers is unlikely, however, with a typical maximum more likely to be about five for a total aggregated bandwidth of about 100 MHz. Bands available in this range have been widely used by 3G and 4G networks.

In the 3.3–4.2 GHz region of FR1, often referred to as the C-Band or the 3.5 GHz band, the operation is TDD, and the maximum spectrum per operator is typically about 100 MHz. This band is likely to be highly used in 5G networks as the amount of spectrum is relatively large yet propagation loss only on the order of about 5 dB more than those bands in the 2 GHz range. Further, this additional path loss can be overcome by utilizing high-gain beamforming antennas leading to coverage similar to that in the 2 GHz bands.

In the 5–6 GHz region of FR1, bands are unlicensed. Such bands are not supported in 3GPP Rel.15 but are expected in the future. They will all offer large amounts of spectrum ranging from about 300 MHz to 800 MHz. Because such bands are available to all, they are accessed on a first come first serve basis, where a potential user must first listen to check if the band is free and only access it if it is. Thus, though the large spectrum potentially affords high data rates, operation is subject to high latency which can lower the average achieved data rate.

In the FR2 range, all operation is TDD, and available spectrum undergoes a large increase relative to FR1. In order of increasing frequency, the first 3GPP specified band covers 24.25–27.5 GHz and is referred to as the 26 GHz band. Next is the

28 GHz band which in fact refers to two bands, namely, one that covers 26.5–29.5 GHz and a second narrower one that covers 27.5–28.35 GHz. Finally, among the specified bands is the 39 GHz band which covers 37.0–40.0 GHz. In these bands, the spectrum per operator range typically up to 400 MHz which supports user rates of up to 5 Gbps. The price paid for these high data rates is much reduced coverage, typically on the order of hundreds of meters or less. Further, this low-coverage problem is made more difficult due to achievable base station output power being generally lower than that in FR1 bands and penetration loss through walls, windows, and doors being much higher than in the lower bands, as discussed in Sect.4.5.6.

Several bands above 40 GHz are under consideration which offer even more spectrum and even higher user data rates, but they each bring their own challenges over and above that of ever-increasing propagation loss. For example, as can be seen in Fig. 4.12, in the 60 GHz region, there is very large oxygen absorption.

Shown in Table 10.4 are operating bands in the 600–800 MHz portion of FR1. It will be seen that the maximum per carrier bandwidth is 20 MHz. Shown in Table 10.5 are those NR operating bands in the mid to high portions of FR1 that afford the largest channel bandwidths per carrier (50 and 100 MHz). These bands are therefore likely to be very desirable. Finally, shown in Table 10.6 are NR operating bands in FR2 where the maximum per carrier bandwidth is 400 MHz.

10.19   5G Data Rates

Extremely high data rates are one of the major features of 5G NR. For NR, the approximate maximum data rate per carrier can be computed as follows [5]:

$$\text{Maximum data rate}(\text{Mb} / \text{s}) = \frac{1}{T_s^{\mu}} \times N_{PRB}^{BW,\mu} \times 12 \times (1 - OH) \times Q$$
$$\times R_{max} \times f \times v \times 10^{-6} \qquad (10.1)$$

where:

- $T_s^{\mu} = {10^{-3}} \big/ {14 \times 2^{\mu}}$ is the average OFDM symbol duration in a subframe of numerology μ, assuming normal cyclic prefix.
- $N_{PRB}^{BW,\propto}$ is the maximum RB allocation in the available system bandwidth BW with numerology μ.
- OH is the time-frequency resource overhead. Simply stated, this is the average ratio of all the resource elements not used by the PDSCH or the PUSCH to the total number of available resource elements. It takes the following values:
- 0.14 for FR1 DL
- 0.18 for FR2 DL
- 0.08 for FR1 UL

Table 10.4 NR operating bands in 600–800 MHz portion of FR1

NR band #	Uplink frequency range (MHz)	Downlink frequency range (MHz)	Duplex mode	Maximum channel bandwidth (MHz)
n12	699–716	729–746	FDD	15
n14	788–798	758–768	FDD	10
n28	703–748	758–803	FDD	40
n71	663–698	617–652	FDD	20

Table 10.5 Selected NR operating bands in mid to high portions of FR1

NR band #	Uplink frequency range (MHz)	Downlink frequency range (MHz)	Duplex mode	Maximum channel bandwidth (MHz)
n7	2500–2570	2620–2690	FDD	50
n40	2300–2400	2300–2400	TDD	80
n41	2496–2690	2496–2690	TDD	100
n48	3350–3700	3350–3700	TDD	100
n50	1432–1517	1432–1517	TDD	80
n77	3300–4200	3300–4200	TDD	100
n78	3300–3800	3300–3800	TDD	100
n79	4400–5000	4400–5000	TDD	100

Table 10.6 NR operating bands in FR2

NR band #	Uplink frequency range (MHz)	Downlink frequency range (MHz)	Duplex mode	Maximum channel bandwidth (MHz)
n257	26,500–29,500	26,500–29,500	TDD	400
n258	24,250–27,500	24,250–27,500	TDD	400
n260	37,000–40,000	37,000–40,000	TDD	400
n261	27,500–28,350	27,500–28,350	TDD	400

- 0.10 for FR2 UL
- Q is the bits per symbol for the applied modulation scheme, being 8 for 256-QAM, the highest-order modulation supported.
- R_{max} is the maximum code rate. In 5G NR, it is 948/1024.
- f is a scaling factor used to reflect the capability mismatch between baseband and RF capability of the UE. It is signaled per band and can take the values 1, 0.8, 0.75, and 0.4.
- v is the maximum of layers.

We now consider why the above equation gives maximum data rate.

- T_s^α: The subframe (slot) duration, as indicated in Sect. 10.7, is given by $\dfrac{10^{-3}}{2^\mu}$ secs. Thus, assuming 14 OFDM symbols per slot, the OFDM symbol duration is given by $\dfrac{10^{-3}}{14 \times 2^\mu}$, and $1/\,T_s^\alpha$ represents the OFDM symbol rate.

- One OFDM symbol is composed of $N_{PRB}^{BW,\mu} \times 12$ modulated subcarriers. Thus, $\frac{1}{T_s^\mu} \times N_{PRB}^{BW,\mu} \times 12$ represents the total modulating *symbol* rate, SR_T say.
- $(1 - OH)$ represents the fraction of REs available for user data transmission. Thus, $SR_T \times (1 - OH)$ represents the fraction of modulating symbols available for user data.
- The user bits per OFDM modulating symbol equals $Q \times R_{max}$. Thus, the user *bit* rate is $SR_T \times (1 - OH) \times Q \times R_{max}$.
- v: Data rate clearly proportional to the number of layers hence the multiplication by v.
- f: This scaling factor allows the introduction of a practical baseband to RF limitation to the calculation.

The rate computed by Eq. 10.1 is for a single-component carrier. For the case of carrier aggregation, then the maximum rate is the sum of the rates of each component carrier. As indicated in Sect. 9.8, NR supports up to 16 component carriers in both FR1 and FR2.

The maximum downlink user data rate attainable with NR with a single-component carrier occurs for:

- A subcarrier spacing of 120 kHz in FR2 with the maximum resource block allocation of 264 (Table 10.1) and hence a maximum bandwidth of approximately 400 MHz
- 8 layers
- R_{max} of 948/1024
- Overhead OH of 0.18
- 256-QAM modulation resulting in 8 bits per modulating symbol
- A scaling factor f of 1

Applying this data to Eq. 10.1 gives a maximum component carrier DL data rate of 17.24 Gb/s.

For the uplink, all the above data remains the same with the exception of the maximum number of layers which is now 4 and the overhead which is now 0.1. Applying this data to Eq. 10.1, we get a maximum component carrier UL data rate of 9.46 Gb/s.

As the maximum number of component carriers supported is 16, then if all were of 400 MHz bandwidth, the maximum DL data rate would be 276 Gb/s!! Such a scenario is highly unlikely, but it does demonstrate the ultimate DL data rate capability of NR. A more likely scenario is two 400 MHz component carriers leading to a maximum DL data rate of 34.5 Gb/s which handily meets the IMT-2020 requirement of a minimum DL peak data rate of 20 Gb/s. Table 10.7 shows maximum NR data rates per layer per component carrier.

All FR2 bands are TDD. Thus, with operation in such bands, the maximum downlink and uplink rates are not mutually exclusive. In typical scenarios, there is more downlink demand than uplink demand. Such a scenario could be, for each slot,

Table 10.7 Maximum NR data rates per layer per component carrier

Frequency range	Subcarrier spacing (kHz)	Bandwidth (MHz)	Downlink rate (Mb/s)	Uplink rate (Mb/s)
FR1	15	20	113	121
FR1	15	50	290	309
FR1	30	100	584	625
FR1	60	100	578	618
FR2	60	200	1080	1180
FR2	120	400	2150	2370

ten ODFM symbols for the downlink, one symbol for transition, and three symbols for the uplink. This would result in peak downlink rates of 71% of the maximum available and peak uplink rates of 21% of maximum available.

It is important to not lose sight of the fact that the data rates shown in Table 10.7 represent the very best achievable and that in the real world will only likely be achieved by a small percentage of UEs. These rates assume that the modulation is 256-QAM and the coding rate the maximum of 984/1024. This will only be the case for high SINR situations and thus likely only for UEs located close to the base station. As UE distance from the base station increases, both the modulation order and coding rate will decrease to the point where at the cell edge the modulation is likely to be QPSK and the coding rate less than maximum. Even if the coding rate stays at the maximum, the peak rate attainable with QPSK would only be one fourth that attainable with 256-QAM. As a result, very roughly speaking, the average maximum data rate per layer attainable by a UE in both the DL and UL will likely be about 50% of the maximum rates shown in Table 10.7.

Clearly, in a multiuser system, where total capacity has to be shared by all users, practical user average data rates will be less than the maximum rates discussed above. In such situations, the average user data rate decreases as the number of users increases but is also dependent on bandwidth, received signal levels, and intercell interference levels.

10.20   Transmitter Output Power and Receiver Reference Sensitivity

There are hundreds of individual specifications that define the parameters of 5G base stations and UEs, all precisely stated in the 3GPP 5G NR technical specifications. Two performance parameters, however, have a large impact on coverage, these being transmitter output power and receiver sensitivity. The difference between the two, plus transmitter and receiver antenna gain, defines the maximum path loss tolerable while maintaining packet error rate above defined minimum. In an ideal world, one thus seeks, within limits, to maximize transmitter output power and minimize the value of receiver sensitivity. This section will review at a high-level NR specified base station and UE transmitter output power and receiver reference sensitivity, the latter being a specified receiver sensitivity in a defined reference channel.

10.20.1   Base Station Transmitter Output Power

3GPP base station specifications apply to wide area base stations, medium-range base stations, and local area base stations, and four classes of base stations are defined [6]:

- BS typed *1–0* and *2–0*: Wide area, medium-range, and local area base stations with a BS to UE minimum distance along the ground of 33, 5, and 2 meters, respectively. Types *1–0* and *2–0* have integrated AASs and operate in FR1 and FR2, respectively.
- BS type *1-C* and *1-H*: Wide area, medium range, and local area base stations with a BS to UE minimum coupling loss of 70, 53, and 45 dB. Both types operate in FR1, with type *1-C* connected to antennas via coaxial cables and type *1-H* having an integrated AAS.

For each base station type, the specified maximum base station output power, per carrier and per antenna, is as follows:

- Types *1-C* and *1-H* Local Area BS: 24 dBm
- Types *1-C* and *1-H* Medium-Range BS: 38 dBm
- Types *1-C* and *1-H* Wide Area BS: No upper limit
- Type *1–0* Local Area BS: 33 dBm
- Type *1–0* Medium Range BS: 47 dBm
- Type *1–0* Wide Area Bs: No upper limit
- Type *2–0*: No upper limit

10.20.2   Base Station Receiver Reference Sensitivity

The reference sensitivity level is the minimum received signal level at which there is a sufficient SINR for the specified reference measurement channel to achieve a throughput that's 95% of the maximum possible.

For base station types *1-C,1-H*, and *1–0*, all of which operate in FR1, reference sensitivity is specified for reference measurement channels with a QPSK modulated signal and a coding rate of 1/3 [6]. Not all sensitivities specified will be shown here, but to give a sense of the values, those for the channel with the lowest total

Table 10.8 Some BS type *1-C, 1-H,* and *1-0* reference sensitivities

Ref. measurement channel	Wide area BS ref. sens. (dBm)	Medium area BS ref. sens. (dBm)	Local area BS ref. sens. (dBm)
G-FR1-A1–1	−101.7	−96.7	−93.7
G-FR1-A1–4	−95.3	−90.3	−87.3

subcarrier bandwidth (4.5 MHz), Channel G-FR1-1-A1–1, and the channel with the highest total subcarrier bandwidth (192 MHz), Channel G-FR-1-A1–4, are indicated in Table 10.8.

For base station type *2–0,* which operates in FR2, reference sensitivity, as with the FR1 base stations, is specified for reference measurement channels with a QPSK modulated signal and a coding rate of 1/3. Here, the specified reference sensitivity is declared by the vendor but must lie within a defined range.

For a wide area BS, the range is −96 to −119 dBm for a reference channel with a 50 MHz bandwidth and shifted upward by 3.15 dB for a reference channel with a 100 MHz bandwidth.

For a medium area BS, the range is −91 to −114 dBm for a reference channel with a 50 MHz bandwidth and shifted upward by 3.15 dB for a reference channel with a 100 MHz bandwidth.

For a local area BS, the range is −86 to −109 dBm for a reference channel with a 50 MHz bandwidth and shifted upward by 3.15 dB for a reference channel with a 100 MHz bandwidth.

10.20.3 UE Transmitter Output Power

For FR1, UE maximum output is specified per power class [7]. Power Class 3, the default power class, covers all bands and specifies a maximum output power of 23 dBm. For bands n41, n77, n78, and n79, Power Class 2 is also applicable, with a maximum output power of 26 dBm.

For FR2, four power classes are specified [8]:

- Power Class 1: Fixed wireless access (FWA)
- Power Class 2: Vehicular UE
- Power Class 3: Handheld UE
- Power Class 4: High-power non-handheld UE

For Power Class 1, maximum output power is 35 dBm.

For Power Classes 2, 3, and 4, maximum output power is 23 dBm.

For all power classes, a maximum EIRP is specified as 20 dB above the maximum output power. This implies a maximum antenna gain of 20 dB.

10.20.4 UE Receiver Reference Sensitivity

In FR1, the UE is required to be equipped with a minimum of two receive antenna ports in certain operating bands and with a minimum of four receive antenna ports in others. Here, the reference sensitivity power level is the minimum mean power applied to each one of the UE antenna ports at which there is a sufficient SINR for the specified reference measurement channel to achieve a throughput that's 95% of the maximum possible. For UEs equipped with two receive antenna ports, reference sensitivity is specified for all operating bands and varying channel bandwidths, with QPSK modulated and 1/6 coding rate reference measurement channels [7]. To give a sense of the sensitivity power levels, the lowest level indicated is −100 dBm and occurs in the lowest channel bandwidth (5 MHz) in certain operating bands. The highest level indicated is −84.7 dBm and occurs in the highest channel bandwidth in a specific operating band.

In FR2, as with FR1, reference sensitivity is specified for reference measurement channels with a QPSK modulated signal and a coding rate of 1/6, with levels given are per power class for the various operating bands and channel bandwidths [8]. The lowest level indicated is −97.5 dBm in Power Class 1 (FWA) for a 50 MHz channel bandwidth. The highest level indicated is −79.3 dBm in Power Class 3 (handheld UE) for a 400 MHz channel bandwidth.

10.21 Key 3GPP 5G NR Physical Layer-Related Technical Specifications (TSs) and Reports (TRs)

This text addresses the 5G NR physical layer key technologies at a level believed to be deep enough to impart a solid understanding but not so deep as to be fully comprehensive. Such an understanding requires strong familiarity with the relevant 3GPP specifications and reports. There are tens of 3GPP 5G NR technical specifications (TS) and technical reports (TR). However, some have a larger bearing on the physical layer than others. As this text is largely focused on the physical layer, a listing of those specifications and reports that largely relate to the physical layer is deemed in order and shown in Table 10.9.

10.22 Summary

In earlier chapters, key 5G physical layer technologies were introduced. In this chapter, where and how in NR these technologies are applied have been shown. The NR physical layer is the most complex and most flexible point-to-multipoint radio-access system introduced to date, offering the potential of unprecedented data rates and low latency. It is largely due to the technologies introduced in earlier chapters of this text and the way in which they have been applied, as shown in this chapter,

Table 10.9 Selected 3GPP technical specifications and technical reports

Spec. #	Title
TS 38.101–1	NR, user equipment (UE) radio transmission and reception, part 1: Range 1 standalone
TS 38.101–2	NR, user equipment (UE) radio transmission and reception, part 2: Range 2 standalone
TS 38.104	NR, base station (BS) radio transmission and reception
TS 38.201	NR, physical layer, general description
TS 38.202	NR, services provided by the physical layer
TS 38.211	NR, physical channels and modulation
TS 38.212	NR, multiplexing and channel coding
TS 38.213	NR, physical layer procedures for control
TS 38.214	NR, physical layer procedures for data
TS 38.215	NR, physical layer measurements
TS 38.300	NR, overall description; Stage-2
TS 38.306	NR, user equipment (UE) radio access capabilities
TR 38.801	Study on new radio access technology: radio access architecture and interfaces
TR 38.202	Study on new radio access technology, physical layer aspects
TR 38.817–01	General aspects for user equipment (UE) radio frequency (RF) for NR
TR 38.817–02	General aspects for base station (BS) radio frequency (RF) for NR
TR 38.912	Study on new radio (NR) access technology

that has made this progress possible. Though 5G NR has been developed primarily for mobile access, it is equally applicable to fixed wireless access. In fact, as indicated in Sects. 10.20.3 and 10.20.4, a UE FWA transmitter power class along with FWA receiver reference sensitivity for this class has been specified for FR2. Though not explicitly indicated, FWA is equally applicable to FR1. In the concluding chapter that follows, we will explore the application of NR to FWA.

References

1. 3GPP TR 38.912 (2018) Study on New Radio access technology, Rel. 15, version 15.0.0
2. Dahlman E et al (2018) 5G NR: the next generation wireless access technology. Academic Press, London
3. Holma H et al (eds) (2020) 5G technology: 3GPP new radio. John Wiley & Sons Ltd., Hoboken
4. Ahmadi S (2019) 5G NR: architecture, technology, implementation, and operation of 3GPP new radio standards. Academic Press, London
5. 3GPP TS 38.306 (2020) NR; User Equipment (UE) radio access capabilities, Rel. 16, version 16.0.0
6. 3GPP TS 38.104 (2020) NR; Base Station (BS) radio transmission and reception, Rel. 16, version 16.3.0
7. 3GPP TS 38.101-1 (2020) NR; User Equipment (UE) radio transmission and reception; Part 1; Range 1 Standalone, Rel.16, version 16.3.0
8. 3GPP TS 38.101-2 (2020) NR; User Equipment (UE) radio transmission and reception; Part 2; Range 2 Standalone, Rel. 16, version 16.3.0

Chapter 11
5G Based Fixed Wireless Access

11.1 Introduction

In Sect 1.3, a brief history of fixed wireless access (FWA) was presented. FWA today implies a relatively high data rate (tens of Mb/s to several Gb/s), two-way wireless connection between fixed user locations and a base station that is connected to a packet data handling core network. The high end of the data rates indicated above are afforded by the use of 5G technology and has resulted in FWA now being considered as a serious contender for broadband services to homes and *small- and medium-sized enterprises* (SMEs), being particularly well suited to situations where limited or no infrastructure exists to provide such services via coaxial cable or fiber. With 5G-enabled FWA, depending on spectrum availability, large chunks of radio spectrum can be utilized to provide consumers with high data rate, low-latency connections. In this chapter, the new FWA opportunity, FWA spectrum, market segmentation, deployment scenarios, remote unit structure, and typical link performance are discussed, reinforcing the key role of 5G technology in FWA viability.

FWA creates the opportunity to 5G *enhanced mobile broadband* (eMBB) providers to significantly increase the impact of their 5G deployment by providing service to two user classes simultaneously, eMBB and FWA. Compared to its wired competitors, such an FWA offering can likely, in many instances, be provided more rapidly and at a lower roll-out cost.

11.2 The New FWA Opportunity

The concept of FWA is not new. The first serious attempt at FWA was the original WiMAX offering in the early 2000s. The short story of WiMAX and equivalent offerings in that time frame was that they all largely failed for two primary reasons. Firstly, they relied upon a completely new stand-alone infrastructure and expensive

© Springer Nature Switzerland AG 2020
D. H. Morais, *Key 5G Physical Layer Technologies*,
https://doi.org/10.1007/978-3-030-51441-9_11

proprietary equipment. Secondly, the data rates offered were unattractive relative to wired offerings which at the time were in a phase of rapid increase in capacity as fiber became more and more prevalent. 3GPP specified 5G FWA, as we shall see below, addresses these two limitations head on.

Rather than requiring new infrastructure, the 5G FWA architecture largely copies the standardized 3GPP 5G eMBB architecture in terms of the RAN and core network. The primary difference is the *customer premise equipment* (CPE) which will normally include a management system. This system provides remote access to the CPE for configuration, performance measurement, and troubleshooting. There are two possible approaches to providing 5G FWA. One would be to set up a purely FWA system. Though technically feasible, it would likely find great difficulty in obtaining an adequate return on investment due to the high cost of the infrastructure relative to the potential revenue. If, however, 5G FWA is offered by operators along with eMBB as depicted in Fig. 11.1, then the equation changes significantly as most of the infrastructure is shared for the two services. Here, the incremental cost to an eMBB operator of offering FWA is relatively small, making an acceptable return more likely.

Low incremental FWA cost is not, however, enough to make FWA attractive. The other factor is user data rate. This point is well made when one considers 4G FWA, which from an architectural point of view is almost identical to 5G FWA. The big difference is user data rate. For the first time, by the use of the greater spectrum availability afforded by 5G, particularly in the mmWave bands, but also the 3.5 GHz band, fiber-like rates are now possible with FWA.

Not only is 5G FWA likely to be competitive with hardwired services in both cost and data rate arenas, but if offered by an existing eMBB provider, then it enables relatively rapid rollout compared to fiber where the time to lay cables in trenches or to string them on utility poles can be long and fraught with unexpected difficulty. Further, in addition to being able to compete with fixed access providers, FWA

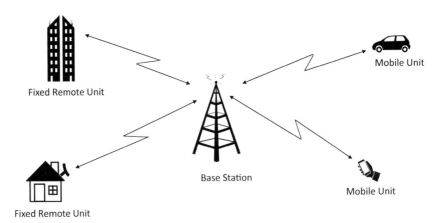

Fig. 11.1 An eMBB and FWA common system

offers the opportunity to mobile providers to profitably serve the unaddressed household market. Shown in Fig. 11.2 is a base station site capable of supporting both mobile and FWA access.

11.3   FWA Suitable Spectrum

In the lower frequency range (FR1), the 600–800 MHz bands provide the best coverage, albeit at measurably lower data rates than higher frequencies. Here all bands are FDD and maximum channel bandwidth varies between 10 and 40 MHz (see Table 10.4), resulting in peak data rates of between about 200 and 800 Mb/s, respectively with 4 x 4 MIMO. Bands n28 and n71, where the maximum channel bandwidth is 20 and 40 MHz, respectively, are the more attractive ones and thus more likely to be deployed. Depending on the number of users per base station, average user data rate in these bands would likely be at least 10 Mb/s.

The FR1 band that offers the best opportunity to provide competitive FWA services, however, is the 3.5 GHz one. Here, typically up to about 100 MHz of TDD spectrum could be available per operator, resulting in a peak data rate of up to about 2 Gb/s with 4 × 4 MIMO and good coverage (up to tens of kilometers with outdoor CPEs).

In the higher frequency range (26, 28, and 39 GHz FR2 bands), typically up to about 800 MHz of TDD spectrum could be available per operator, resulting in a peak data rate of up to about 7.5 GHz with 2 × 2 MIMO, albeit at the expense of coverage (up to about 10 kilometers with outdoor CPEs). Relative to mobile broadband access, FWA coverage and data rates will always have the option to be

Fig. 11.2 A base station site that's capable of supporting mobile and FWA access. (Courtesy of Ericsson AB)

better by using, on the CPEs, high-gain outdoor antennas and, in the case of FR2 bands, higher transmitter output power (Sect. 10.20.3).

Table 11.1 summarizes the coverage, typical spectrum per operator, and typical peak user data rates likely for 5G FWA frequency bands. Bear in mind, however, as discussed in Sect. 10.19, that average user data rate is likely to be measurable less than these peak rates.

11.4 FWA Market Segmentation

Market segmentation is not a science and thus there are many ways to address it. One approach is that offered by Ericsson, where the market is viewed as being divided into three segments, namely, Wireless Fiber, Build with Precision, and Connect the Unconnected [1].

The *Wireless Fiber* segment encompasses those situations where there is a need for very high data rate offerings and cell capacity. Typical data rates in this segment would be in the range 100 Mb/s to 1+ Gb/s. Here FWA is offered as a direct alternative to high-end fixed broadband. The goal is to provide fiber comparable speed and thus handle typical household TV HD needs and SME needs, this coupled with a corresponding high ability to afford the service. The most likely geographic location for this service is urban edges and suburban areas.

The *Build with Precision* segment consists of those situations where the competition is performance-limited broadband alternatives, such as xDSL. Here the need is for higher data rates than available, typical FWA offered rates being in the range 50 to 200 Mb/s. The most likely geographic locations for this service are suburban areas and rural villages and towns that are currently underserved. Note that a downlink rate of 50 Mb/s or more easily supports simultaneous streaming to two 4 K Ultra HD televisions (15 to 25 Mb/s each) plus other relatively high rate data downloads.

The *Connect the Unconnected* segment consists of those situations where broadband competition is virtually nonexistent, and smartphones that use mobile broadband are the dominant way of accessing the Internet. User expectation of accessed

Table 11.1 Likely FWA frequency bands

Band	Coverage	Typical spectrum per operator	Typical peak user data rate
600–800 MHz	Very good: up to tens of km with indoor CPE	20–40 MHz, FDD	400–800 Mb/s with 4 × 4 MIMO
3.5 GHz	Good: up to tens of km with outdoor CPE	100 MHz, TDD	Good: 2 Gb/s with 4 × 4 MIMO
26, 28, & 39 GHz	Limited: up to about 10 km with roof mounted CPE	800 MHz, TDD	Very good: > 7 Gb/s with 2 × 2 MIMO

data speed is low, with typical FWA offered rates being in the range 10 to 100 Mb/s. Even a low download rate of 10 Mb/s, however, still supports 1080p HD television. Likely geographical locations for this service are unserved rural villages. Systems operating in frequency bands in the 600–800 MHz range provide a good solution for these situations as infrastructure cost is relatively low, given that coverage per base station is very good, and achievable data rates are acceptable.

11.5 FR2 Deployment Scenarios

Two common FWA FR2 deployment scenarios are a suburban one and an urban landscape one and are depicted in Fig. 11.3.

In the suburban deployment, which would fall into the Wireless Fiber segment, the goal is to deliver high data rates to homes and SMEs. Tower heights would typically be in the 15 to 25 m range and cell radius in the 500 to 1000 m range, with average house roof height of about 10 m. At the remote end, the customer premise equipment (CPE) would be mounted outdoors, either on the roof or at the top of a high wall. For such a scenario, the base station transmitted vertical beam spread would typically be about 6 to 12 degrees, created by combined array columns of 16 to 8 elements each, respectively, and as the terminals are fixed, there would be no need for fully adaptive elevation scanning. The elevation beam pattern would be directed in an overall downward direction and its EIRP designed such as to increase at approximately the same rate as the path loss, providing a relatively uniform coverage for both near and far users. In the azimuth, a 13 to 26 degree beamwidth

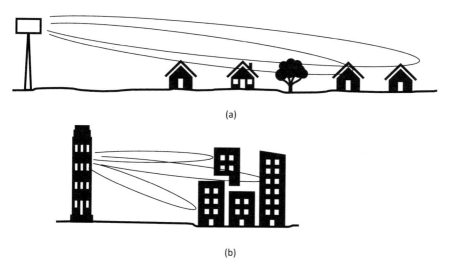

(a)

(b)

Fig. 11.3 Two common FWA FR2 FWA deployment scenarios. (**a**) Suburban. (**b**) Urban

would likely be sufficient and afforded by passively combined array rows of 8 to 4 elements each, respectively.

In the urban deployment, which would also fall in the Wireless Fiber segment, the base station would be mounted lower to the ground than in the suburban deployment and likely on the top or side of a building. Here the base station would likely need vertical scanning capability in order to direct signals across the entire building elevation. Horizontal scanning capability would also be likely needed. The CPEs would likely need to be indoors, thus incurring outdoor-to-indoor penetration loss.

11.6 Customer Premise Equipment Types

Customer premise equipment (CPE) is a key component of the FWA offering, being the main physically distinguishing feature relative to mobile broadband and impacting directly end-user achievable data rates and quality. However, despite its differences with standard smartphones, it still benefits from some high production volume chipsets used in smartphones, thus controlling the total cost. Broadly speaking, there are two types of CPEs, namely, outdoor-mounted and indoor-mounted. Within these types, however, there are variations. Following is some detail on these two types which has been derived in part from [2].

The outdoor CPE can be wall-mounted or roof-mounted, directed to the best serving cell, and provides the best performance. This is because it includes a directional antenna giving increased gain relative to indoor CPE units (e.g., 10–14 dBi at 3.5 GHz) and can thus be installed with a BS to CPE predictable link quality, given that it is not subject to movement as in the case of mobile broadband. The transmission mode (number of layers) for a single outdoor CPE is typically rank-2 (2×2 MIMO) for transmission in FR2 bands as it is expected that the modem will be installed with good line of sight or near line of sight. For transmission in FR1 bands, the transmission mode is typically rank-4 (4×4 MIMO). In terms of spectrum efficiency, the outdoor CPE is typically two to three times more efficient than the indoor type. The outdoor CPE consists of the antenna, RF front end, and RF to baseband functionality and is connected via an Ethernet cable to an indoor router which facilitates end-user connection over Wi-Fi or Ethernet. The outdoor unit typically obtains power via the Ethernet connecting cable, thus eliminating the need for a second cable. Outdoor CPE units are normally owned, controlled, and managed by the service provider.

Indoor CPE units contain the necessary RF front end, RF to baseband, and router functionality and are thus a one-box solution that's easy to install and connect to the network. Its main disadvantage relative to the outdoor type is its much lower antenna gain, making it less suitable in situations where propagation and outdoor-to-indoor loss is high or where many households are served as it utilizes more hertz of RF bandwidth per data bits per second. It typically has two receiver antennas but can also be offered with four to provide improved performance. Unlike outdoor CPE units, indoor CPE units are normally customer-owned, being sold in the operator's retail stores.

The largest difference between outdoor and indoor CPE types is the ability to accomplish the service levels promised, particularly during the busy hours. In terms of the radio resources required, an indoor CPE is comparable with or slightly worse than a smartphone, given that it is always located indoors. By contrast, an outdoor CPE unit enjoys an advantage of approximately 20–30 dB better signal quality, the result of a 10–15 dB difference in antenna gain coupled with the avoidance of another 10–15 dB in outdoor-to-indoor losses. This advantage results in higher data rates and better coverage, all factors that are particularly valuable toward the cell edges in 3.5 GHz and millimeter wave deployments.

To get a sense of the highly integrated technology currently being deployed in support of 5G CPEs, it is instructive to take a high-level look at two modules offered by Qualcomm Technologies to OEM providers of fully developed CPEs. One module, developed exclusively for 5G FWA applications in FR2 bands, is the QTM527 mmWave antenna module which comprises a transceiver (upconverter/downconverter), RF front end (transmitter output power amplifier, receiver LNA, etc.), and an array antenna. It is designed to be paired with Qualcomm Snapdragon X55 5G Modem-RF System which comprises the modem. Together, the two modules thus create a complete mmWave modem to antenna system.

A few key features of the QTM527 are:

- A 64-element dual-polarized antenna capable of a transmitted EIRP of over 40 dBm with Power Class 1 devices
- Support of up to 800 MHz of bandwidth in the 26, 28, and 39 GHz FR2 bands
- Support for up to 2 × 2 MIMO with dual-layer polarization in both the downlink and uplink
- Support for beamforming, beam steering, and beam tracking for bidirectional communications

The Snapdragon X55, which is designed to address both the mobile and FWA markets, supports the QTM527 mmWave bands, with a capability in these bands of 7.5 Gb/s peak downlink data rate and 3 Gb/s peak uplink data rate (as transmission is TDD, these rates are not mutually exclusive). It also provides an interface to Sub-6 GHz band transceivers, supporting up to 200 MHz of bandwidth and 4 × 4 MIMO, thus implying peak data rates of up to about 4 GB/s. This interface allows OEMs to add circuitry to cover, for example, the 3.5 GHz band and/or the 600 and 700 MHz bands.

Shown in Fig. 11.4 is the structure of a typical outdoor-mounted CPE indicating the indoor unit, the Ethernet cable connecting the indoor unit to the outdoor unit, and the outdoor unit which, in the case depicted, can provide connectivity via both Sub-6 GHz bands and mmWave bands.

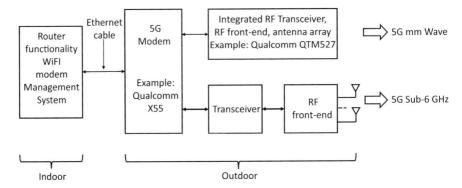

Fig. 11.4 Typical outdoor CPE structure

11.7   Two 5G FWA System Performance Studies

A number of studies regarding the likely performance characteristics of 5G FWA systems have been carried out. Among these are one by Ericsson researchers and another by Nokia Bell Labs researchers. Summaries of their findings are presented below as they provide good insight into likely 5G FWA performance.

11.7.1   An Ericsson Performance Study

Ericsson carried out a study of a possible 5G FWA deployment and published its findings in the Ericsson Technology Review [3]. This study and its findings are very informative and are summarized below.

The key parameters of the simulated system are:

- Suburban environment with 1000 households per square kilometer.
- 25% of households use a 4 K HD video service requiring a download rate of ≥15 Mb/s.
- Base stations mounted on utility poles 6 m tall.
- CPE antennas placed outdoors as well as indoors.
- CPE buildings 4–10 m tall.
- Trees in the neighborhood 5–15 m and attenuate the signal.
- 35 dBm base station transmitter power.
- 30 dBm CPE transmitter power.
- 200 MHz channel bandwidth.
- 28 GHz operating frequency.
- TDD duplex mode, with 57% DL allocation.
- Beam forming and multiuser MIMO.
- Base station antenna array of 8 × 12 cross polarized elements.
- Omnidirectional CPE antennas with a 10 dBi gain.

– Two-layer MIMO for each user.

The simulation results indicated the following:

– A majority of the users enjoy data rates in excess 800 Mb/s.
– Only 11 percent of users have a data rate below 400 MHz, and all meet the target of ≥15 Mb/s.
– As traffic load increased, user data rate decreases due to greater interference and queuing. For example, for CPE units with rooftop antennas and for user traffic volume (GB/subscriber/month) of 3000, 100% of users have a data rate > 15 Mb/s, and 93% have a data rate > 100 Mb/s. However, for a user traffic volume of 5200, 95% of users have a data rate > 15 Mb/s, but only 69% have a data rate > 100 Mb/s.
– For intercell site distance (ISD) of 350 m, 78% of users could use indoor antennas, 17% could use outdoor wall-mounted antennas, and only 5% required roof-mounted antennas.
– A similar analysis was done on the 3.5 GHz band, and Fig. 11.5 shows the difference in coverage based on operating frequency, CPE antenna location, and foliage. We note that 3.5 GHz provides, as expected, better coverage. This, however, is likely to be at the expense of data rate and capacity given that less RF spectrum is likely available.
– Coverage at 28 GHz was shown to be strongly dependent on foliage. Deploying base station antennas at a height greater than that of the tallest trees in the coverage area significantly increased the cell range.

11.7.2 A Nokia Bell Labs Performance Study

In [4], simulation results of various performance characteristics of 5G mmWave FWA systems are presented. Among these are the effect of ISD, foliage, and indoor versus outdoor CPE mounting on the mean and cell-edge user data rates. A study of

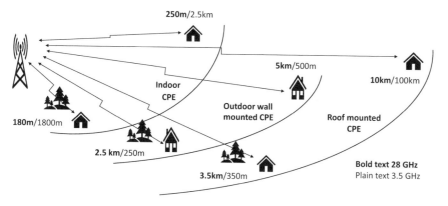

Fig. 11.5 Coverage at 3.5 GHz and 28 GHz as a function of CPE antenna location and foliage

these varying rates is helpful in understanding likely 5G mmWave FWA performance. The key parameters of the simulated systems will thus be presented followed by some of the results.

The key parameters of the simulated system are:

- 28 GHz operating frequency
- 800 MHz RF bandwidth via the aggregation of 8 100 MHz carriers
- TDD with 50% DL/50% UL split
- Base station: 6–8 m height, three-sector antenna
- Base station antenna array: 512 elements each. Can be single-panel or four-panel
- CPE antenna system: Single two rows by 4-column panel array of cross-polarized antennas, resulting in 16 elements
- Base station transmitter power: 16 dBm per antenna element, 43 dBm total power (20 Watts)
- CPE transmitter power: 11 dBm per element, 23 dBm total power
- Maximum rank (SU MIMO): Two for single-panel array, eight for four-panel array
- Maximum number of CPE user streams for MU MIMO: 4
- CPE mounting: outdoor or indoor

Some of the simulation results are shown in Figs. 11.6, 11.7, 11.8, 11.9 and indicate the effect of some important parameters on downlink and uplink data rate performance. Specifically, they show data rate performance for ISD values 100 and 300 meters, with no foliage and heavy foliage and with CPE units mounted outdoors or indoors. Some summary observations from the simulations are as follows:

- With no foliage, an average DL data rate in excess of 400 Mb/s and a cell-edge data rate in excess of 100 Mb/s can be achieved with both outdoor- and indoor-mounted CPEs for an ISD distance of 300 m.

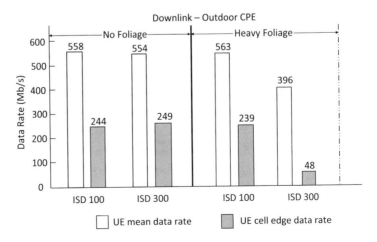

Fig. 11.6 Effect of ISD and foliage on outdoor CPE DL mean and cell-edge data rate

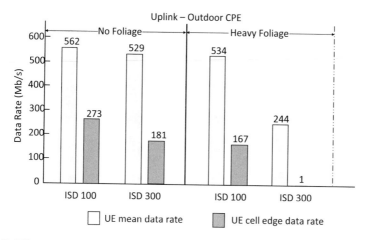

Fig. 11.7 Effect of ISD and foliage on outdoor CPE UL mean and cell-edge data rate

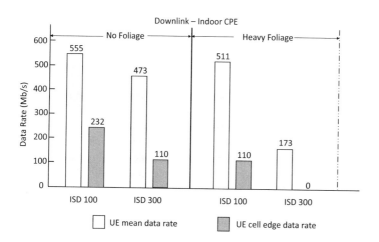

Fig. 11.8 Effect of ISD and foliage on indoor CPE DL mean and cell-edge data rate

– With heavy foliage, an average DL data rate in excess of 400 Mb/s and a cell-edge data rate in excess of 100 Mb/s can be achieved with both outdoor- and indoor-mounted CPEs for an ISD distance of 100 m.
– With no foliage, an average UL data rate in excess of 400 Mb/s and a cell-edge data rate in excess of 100 Mb/s can be achieved with both outdoor- and indoor-mounted CPEs for an ISD distance of 100 m.
– With heavy foliage, an average UL data rate in excess of 400 Mb/s and a cell-edge data rate in excess of 100 Mb/s can only be achieved with outdoor-mounted CPEs for an ISD distance of 100 m.

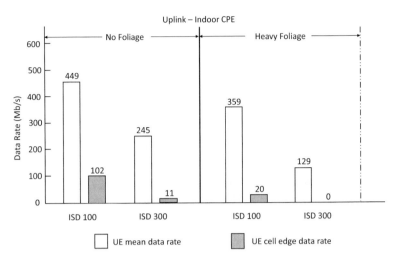

Fig. 11.9   Effect of ISD and foliage on indoor CPE UL mean and cell-edge data rate

11.8   Conclusion

In Chap. 1, mobile and fixed wireless access was introduced at a high level. As such access takes place in a cellular environment, cellular coverage methods were discussed. In Chap. 2 the technical nature of broadband wireless payload was reviewed to aid later in understanding the data being processed in the 5G NR Radio Access Network. Chapter 3 provided some mathematical tools to aid in the understanding of technical concepts that follow. Chapter 4 outlined the nature of the wireless path traversed by the radio signals in a cellular environment. The information provided in Chaps. 1, 2, 3, 4, though peripheral, was important to truly understand the need for, and the necessary capabilities of, 5G NR key technologies. Chapters 5, 6, 7, 8, 9 addressed these technologies. Information was provided at a level sufficient to impart a fundamental grasp of the structure and functioning of these technology components, but not at a level so deep as to make it somewhat intractable to those with a limited background in the subject. In Chap. 10, many details of the 5G NR physical layer were outlined, indicating clearly the roll of the technologies earlier discussed in making the features and performance of the 5G NR physical layer possible. Finally, in this chapter, fixed wireless access was introduced, showing the symbiotic relationship between mobile broadband and FWA in a 5G NR environment.

Mobile broadband is an ever-changing technology, striving to always improve service. This improvement comes, however, at the expense of ever more complex technology. There is no reason to believe that many as yet undiscovered technologies will not emerge and supplement or replace those in current use. That said, a lot of technologies presented in this text will likely have a long shelf life. Take digital modulation. Many variations exist, but the fundamental property of allowing an increase in spectral efficiency at the expense of poorer signal-to-noise characteristics is unchanging. The same long shelf life comment applies equally to channel coding. Yes, it's true that no sooner is a particular type of coding crowned as the ultimate that a new one emerges. In the last decade, turbo convolution coding has been supplanted by low-density parity check coding which, at least for certain applications, now has polar coding nipping at its heels. All is true, but the fundamental need for error correction via coding schemes will be a necessity for the foreseeable future. It is thus hoped that, having studied this introduction to the key physical layer technologies of 5G NR and the 5G NR physical layer itself, should the reader desire to explore this subject area in greater detail, he/she will feel well-positioned to do so.

Acknowledgement

The author would like to thank Sven Hellsten of Ericsson for his valuable help during the preperation of this chapter.

References

1. Olofsson H et al (2018) Leveraging LTE and 5G NR networks for fixed wireless access. Ericsson Technology Review, Aug 2018
2. Ericsson AB (2019) Fixed wireless access handbook
3. Laraqui K et al (2016) Fixed Wireless Access on a massive scale with 5G. Ericsson Technology Review, Dec 2016
4. Vook F et al (2016) Performance characteristics of 5G mmWave wireless-to-the-home. In: IEEE 50th Asilomar Conference on Signals, Systems and Computers, Pacific Grove, Nov 2016, pp 1181–1185

Correction to: Key 5G Physical Layer Technologies

Douglas H. Morais

Correction to:
D. H. Morais, *Key 5G Physical Layer Technologies*,
https://doi.org/10.1007/978-3-030-51441-9

The book was inadvertently published without updating the following corrections:

Corrections:

P. 15: Figure 2.2 was published incorrectly in the original version of the book. The correct version of the figure is updated in the online version of this chapter.

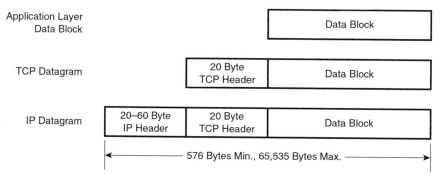

Fig. 2.2 TCP and IP encapsulation

The updated online versions of these chapters can be found at
https://doi.org/10.1007/978-3-030-51441-9_2
https://doi.org/10.1007/978-3-030-51441-9_6
https://doi.org/10.1007/978-3-030-51441-9_7
https://doi.org/10.1007/978-3-030-51441-9

P. 16, Line 10 from bottom: The text was processed incorrectly in the original version of the book. The text should read as below:

"The maximum size of a TCP-derived IP datagram is 65,535 bytes".

P. 17, Line 12: The text was processed incorrectly in the original version of the book. The text should read as below:

"eliminated by using IP datagrams no longer than 576 bytes, since all IP…".

P. 104: The presentation of the division processed incorrectly. The corrected presentation of the division is given in the updated online version of this chapter.

P. 147: The caption for Figure 7.13 was processed incorrectly. The correct figure caption is listed below:

Fig. 7.13 Phase noise PSD as per Leeson. Lower, middle, and upper traces are 2, 10, and 30 GHz respectively. (From [8], with permission of Elsevier)

Appendix: Helpful Mathematical Identities

Trigonometric Identities

$$\sin(x \pm y) = \sin x \cos y \pm \cos x \sin y$$

$$\cos(x \pm y) = \cos x \cos y \pm \sin x \sin y$$

$$\sin x \sin y = \frac{1}{2}\cos(x-y) - \frac{1}{2}\cos(x+y)$$

$$\cos x \cos y = \frac{1}{2}\cos(x+y) + \frac{1}{2}\cos(x-y)$$

$$\sin x \cos y = \frac{1}{2}\sin(x+y) + \frac{1}{2}\sin(x-y)$$

$$\cos x \sin y = \frac{1}{2}\sin(x+y) - \frac{1}{2}\sin(x-y)$$

$$\sin^2 x = \frac{1}{2}(1 - \cos 2x)$$

$$\cos^2 x = \frac{1}{2}(1 + \cos 2x)$$

$$\sin x = \frac{e^{jx} - e^{-jx}}{2j}$$

© Springer Nature Switzerland AG 2020
D. H. Morais, *Key 5G Physical Layer Technologies*,
https://doi.org/10.1007/978-3-030-51441-9

$$\cos x = \frac{e^{jx} + e^{-jx}}{2}$$

$$e^{jx} = \cos x + j \sin x$$

Standard Integrals

Where a, b, and c are constants,

$$\int \sin(ax+b)\,dx = -\frac{1}{a}\cos(ax+b) + c$$

$$\int \cos(ax+b)\,dx = \frac{1}{a}\sin(ax+b) + c$$

$$\int a\,dx = ax + b$$

$$\int (ax+b)^n \, dx = \frac{1}{a(n+1)}(ax+b)^{n+1} \quad n \neq -1$$

$$\int e^{ax+b}\,dx = \frac{1}{a}e^{ax+b} + c$$

Matrix Algebra

Matrix Product Example

$$
\underset{3\times 2}{\begin{pmatrix} a_{11} & a_{12} \\ a_{21} & a_{22} \\ a_{31} & a_{32} \end{pmatrix}}
\underset{2\times 2}{\begin{pmatrix} b_{11} & b_{12} \\ b_{21} & b_{22} \end{pmatrix}}
= \underset{3\times 2}{\begin{pmatrix} a_{11}b_{11}+a_{12}b_{21} & a_{11}b_{12}+a_{12}b_{22} \\ a_{21}b_{11}+a_{22}b_{21} & a_{21}b_{12}+a_{22}b_{22} \\ a_{31}b_{11}+a_{32}b_{21} & a_{31}b_{12}+a_{32}b_{22} \end{pmatrix}}
$$

Index

© Springer Nature Switzerland AG 2020
D. H. Morais, *Key 5G Physical Layer Technologies*,
https://doi.org/10.1007/978-3-030-51441-9

Printed in the United States
by Baker & Taylor Publisher Services